WINDFARM VISUALISATION

PERSPECTIVE OR PERCEPTION?

ALAN MACDONALD RIBA

Whittles Publishing

Published by
Whittles Publishing,
Dunbeath,
Caithness KW6 6EG,
Scotland, UK

www.whittlespublishing.com

© 2012 Alan Macdonald

ISBN 978-184995-053-4

The publisher and authors have used their best efforts in preparing this book, but assume no responsibility
for any injury and/or damage to persons or property from the use or implementation of any methods,
instructions, ideas or materials contained within this book. All operations should be undertaken in
accordance with existing legislation and recognized trade practice. Whilst the information and advice
in this book is believed to be true and accurate at the time of going to press, the authors and publisher
accept no legal responsibility or liability for errors or omissions that may have been made.

Printed and bound in Finidr

This book is dedicated to Diana, in appreciation of her many years of experience in the field of research, and Bruce, who was responsible for most of the graphics and the layout of this book. Without their respective talents, this publication would not have been possible.

For up-to-date information, please refer to www.windfarmvisualisation.com

CONTENTS

ACKNOWLEDGEMENTS

I would like to thank The Highland Council Planning and Development Service for their co-operation, in particular their former Principal Planning Officer, Gordon Mooney, for his determination, help and encouragement in researching this book. It is a very rare confluence to meet an experienced planner who is also a photographer and a proficient mathematician with a long-standing interest in cognitive science, to say nothing of his renown as a piper and instrument maker. I am greatly indebted to him for his scientific input and for providing me with his personal research material.

I also acknowledge the help and input I have received from Perth and Kinross Council and in particular Graham Esson, their Team Leader of Strategic Policy and Sustainable Development.

My special appreciation goes to Professor David Knill of the Center for Visual Science, Rochester, USA, who has answered my questions with great patience and kindly vetted my original material relating to human vision and cognitive science. I am also grateful to Richard Burden and Professor Patrick Devine-Wright for their constructive comments on my draft manuscript.

I would also like to express gratitude to my many friends and associates who have tested images and posed for photographs around which many of the graphics in the book have been created, in particular, my friends Rowland and Clare, who have sustained and endured this project from its inception.

My thanks also to Bridgeman Art, Early Photography, Josef Pauli and The Keil Cognitive Systems Group, Oxford University Press, University of Stirling, Frances Lincoln Ltd, and artists Mike Taylor and I-Ming, for the use of their images and work in this book. All other graphics, photographs and photomontages have been produced by Architech Animation Studios in Inverness, Scotland. I also acknowledge the kind permission of Kimbolton School, Stop Bicton Wind Farm, Historic Scotland, Save Our Silton, Combined Community Council Group (Druim Ba) and The Highland Council for their permission to include photomontages commissioned by them.

Finally, I would like to thank the many communities and individuals that I have talked to over the last decade on this issue. From Cornwall to the Butt of Lewis, countless people have expressed their opinions on this subject and provided information and welcome support.

The outcome of any book project is greatly influenced by the nature and standards of the publisher. I am exceedingly fortunate to have found in Whittles Publishing, a technical publisher who not only appreciated the need for this specialist subject to be more widely understood, but have also steered me through the rather daunting process with such a light and courteous touch. Their many suggestions have considerably improved the text and their high production values have proved more than equal to the demands of such highly visual material.

No acknowledgement would be complete without recognition of the work of the late Professor John Benson. I am beholden to him for his clear insights into the problems which have provided a primary basis for my research and a continual source of encouragement in moments when the smokescreen of complexity seemed impenetrable.

PREFACE

In a critique of my original manuscript for this book, an eminent person who had read my material constructively recommended that I should put my credentials up front and explain why I thought I was the best person to write a book on this subject: a suggestion which required considered thought and one I can hopefully answer, but I respectfully acknowledge that the reader will be the ultimate judge of that.

The difference between a photograph and what we see in reality first raised my curiosity following a month's trip to India and Nepal in the mid-1980s. To record my experiences, I was equipped with a Canon 35mm camera fitted with a standard 50mm lens, a 35mm wide-angle lens and many rolls of slide and negative film. When I returned to Hong Kong where I was working at the time, I eagerly awaited the photographic results of my adventure. They did indeed capture the atmosphere and colour of what I experienced, with one exception: the images that I had taken of the Himalayas were so disappointing because they did not reflect the grandeur and scale of what I had seen. The 50mm images somehow made the mountains appear further away than I remembered and the images taken with the 35mm wide-angle lens which showed the mountains in a wider context were even worse.

The printed images were particularly disappointing, but I found that if the 50mm slides were projected onto a white screen at a larger size they began to appear more realistic, and the bigger I scaled the image by moving the projector closer to the screen, the nearer it matched the memory of my experience. In camera terms, I was simply increasing the focal length to make the mountains look nearer. What I could not understand at the time was if a 50mm lens was universally accepted as the lens which best represented what is seen by the human eye, why did this under-representation occur? Little did I realise that this same question would become such an important issue over a quarter of a century later.

When I set up my visualisation company in Hong Kong in 1992, the development of photo-realistic 3D computer software was in its infancy. However, there were one or two advanced packages which incorporated accurate computer cameras which could be set to the same focal length of the actual film camera which took the photograph. This enabled a computer graphic of any proposed development to be accurately matched to a photograph of a known focal length and location. There were no guidelines available at the time for photomontage submitted with planning applications because it was an emerging technology, but potentially a very powerful one, given the old adage that a picture speaks a thousand words.

The coming together of two disciplines, computer science and photography, enabled the production of photo-realistic visualisations which could accurately show a future development superimposed onto a real photograph. The advantages were instantaneous: different designs could be easily tested, the developments could be shown in their real context and the visual impact of complex developments could be clearly understood by decision makers. It was a revolutionary new technology which could be used to both inform the planning system and promote better design. In Hong Kong, the technique was quickly embraced by the planning system and become mandatory for all large developments. In the UK, however, photomontage visualisation is still not a mandatory part of the general planning system, although it is now an accepted part of Environmental

Impact Assessment, with some alarming and perverse results. Photomontage techniques submitted in windfarm applications are being used to deceive rather than inform, justified by a highly complex viewing technique which is fundamentally flawed. This book explains why.

I have also given much consideration to how this book should be written. On the one hand, I could have written it in a more technical way which might have more academic standing, but I have decided to write it as a straightforward story based on my personal experience and knowledge, which should, in my view, go hand in hand. It is the only way to explain my long journey into such a wide range of fascinating disciplines which impinge on our understanding of this subject. A degree of technical discussion is necessary because of the complexity which has been imposed on something so simple and because many of the arguments put forward by windfarm applicants and their consultants have no substance. However, with the help of many illustrations, I hope that by the end of the book the reader will not only be more aware of the problems with the visualisations and the possible solutions, but that they will also be more aware of how we perceive the world around us using our remarkable gift of sight – something most of us never give a second thought to.

Alan Macdonald RIBA

SECTION 1

1

INTRODUCTION

PHOTOMONTAGE AND SCALE

Photomontage visualisation is simply a technique of combining a photographic background with a computer model of a proposed development in order to provide an accessible visual prediction. Architects' plans are sometimes not readily understood by the general public and while the artistic illustrations used in the past were helpful, they could never attain the same level of reality as photography.

Current computer technology enables any building or infrastructure project to be visualised in considerable detail and theoretically should preclude future planning mistakes and misjudgement. Over the years I have attended many planning presentations and have observed on every occasion the attention given to visualisations in preference to all other documentation. We live in a world in which imagery predominates and the importance and accessibility of informative visual material should not be underestimated. However, we should be aware that this same technology also provides the opportunity to manipulate imagery as never before.

Photomontages are created for a number of different reasons. They may be created as an artistic impression for promotional purposes, designed to show how a development will relate to its wider surroundings or to give a realistic impression of perceived scale from an actual viewpoint. If, however, they are to be used within the planning system, the purpose of the montage should be clearly stated and the scale of a potential development should be easily understood without the possibility of any misinterpretation. It is simply a matter of experience and straightforward common sense to provide the viewer with sufficient information.

Fig. 1. If we look at the image above, we can see that one box is half the size of the other, but what size are they?

Fig. 2. If we now add a third element which is familiar, we instantly get a sense of scale and an idea of their size.

Building developments are generally shown in the foreground of a photomontage. If we look at the two images below, we can easily form a sense of scale of the three new houses in the centre of the image, even though the photograph was taken with a wide-angle lens. This is because there are many elements which are familiar to us in our everyday environment.

Fig. 3. Original wide-angle photograph of development site.

Fig. 4. Photomontage of proposed housing development.

The original photomontage in 1996 of Cesar Pelli's IFC2 Tower shown in Fig. 5 below also contained enough visual information for the Hong Kong public to form a sense of scale. Although it might take some time for someone elsewhere to assess its height because structures of this scale are generally unfamiliar to us, people in Hong Kong could quickly form a sense of scale because they are constantly surrounded by tall structures and are familiar with the many landmark buildings in the photograph which they can use for scale reference.

Fig. 5. The original photomontage of the IFC2 Tower, the tallest building on Hong Kong Island, shows the scale of the development in relation to the existing urban fabric.

A photomontage of a building located on the corner site of a busy street intersection taken from the diagonally opposite corner can also be taken with a wide-angle lens, although in reality we will only see a small part of the building within our clear vision at any given time. We can assess its potential scale by the number of floors and by comparing its height to the adjacent buildings in the street, along with other familiar cues such as cars and people. The question of focal lengths, however, becomes more problematic in photomontages specifically commissioned for visual impact assessment if the development is some considerable distance away, and this is particularly important in the case of windfarm visualisations.

Fig. 6.

In the above image, what height are the turbines? In relation to the house in the middle foreground, are they 70 metres high and just beyond the trees, or are they 125 metres high and much further away? We have no way of accurately assessing their height because, unlike buildings, they do not have any recognisable scaling features, due to their simplistic form. If the turbines were replaced by multi-storey residential blocks of the same height, we would be able to form an impression of scale and even an idea of distance relative to the foreground house, because we know from experience that a tower block's storey height is approximately the same as the ground floor of a house.

In reality, if the turbines were built, the situation would be different. Our ability to assess distance within the 3D world which surrounds us would in turn give us a sense of scale and this would become even more apparent as we move through the landscape, because more distant objects appear to move more slowly. Because wide-angle lenses diminish the scale of more distant objects and the further their distance the greater the problem, the choice of focal length becomes important. The only way we can provide the viewer with a sense of how big the wind turbines will appear from the viewpoint is to provide an image with a focal length which is similar to what we see.

To understand the effect of different focal lengths on our perception of distance and scale, look at the two photographs below. Against the same landscape background which contains an existing windfarm, the images were taken from two different camera locations so that the person in the foreground who is standing in exactly the same position appears the same size in both images. In the left-hand image taken with a 24mm wide-angle lens, the turbines in the landscape behind appear much further away and are barely visible compared to the turbines in the right-hand image which was taken with a 105mm telephoto lens. In reality it can be seen by the landscape features that the turbines are the same size.

Fig. 7.

Fig. 8.

THE DEVELOPMENT OF WINDFARM VISUALISATIONS

Up until the mid-1990s, the visuals in the environmental statements were relatively simple, straightforward and easily understood by everyone, although far from ideal. The presentation format was generally A4 size with 50mm single frame montages to show the visual impact, along with fold-out A3 size pages containing panoramic views made up of overlapping photographs to show the wider landscape's character and context.

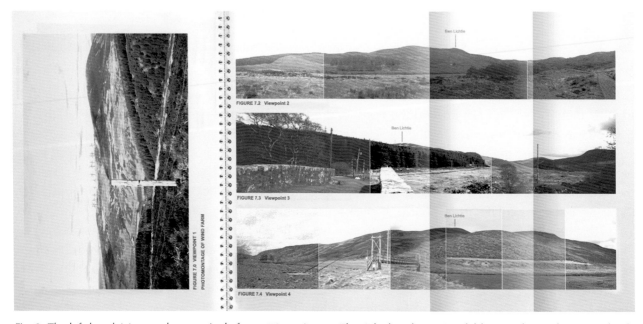

Fig. 9. The left-hand A4 page shows a single frame 50mm image. The right-hand page is a fold-in A3 sheet where it is clearly evident that the panoramic images are made up of several single frame photographs spliced together horizontally.

Because the panoramic images were made up of photographs which were butt-jointed with the joins clearly visible, it was obvious to any viewer that the image was made up of many different photographs and that the angle of view was much wider than what we see in reality within our clear vision at any one time. The separate single frame images, rather crudely montaged as they were, did give a reasonable representation of the landscape scale, although the quality of the photography and photocopying was poor and the turbines hardly visible.

The confusion for the public, non-landscape professionals and decision makers began with the availability of computer software which could join photographic images together to form a seamless panoramic image. Because there were no visible joins, it was easy to assume that it was a photograph which had been cropped top and bottom.

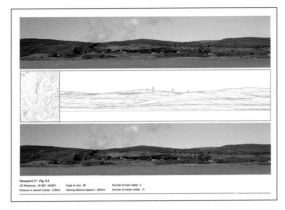

Fig. 10. Three images per A3 page.

Fig. 11. Two images per A3 page.

Fig. 12. Extended A3 panoramas up to A1 width.

The additional single frame image where the visual impact was more easily understood was replaced by a methodology involving a viewing distance, applied to the stitched panorama. The 50mm photograph which was reduced in size now only formed the central portion of a much wider image so it could be viewed in correct perspective with additional peripheral vision. In effect, one image was now being used for both visual impact assessment and landscape assessment.

From the outset, the wide panoramas attracted claims of misleading representation. Initially, the viewing distance was not stated and when it was added to the burgeoning technical data, there was no explanation of what it was or how to apply it. By the late 1990s it was becoming an issue at public inquiries and any criticism or concerns raised by the public were rejected on the grounds that the visualisations were constructed in accordance with best practice.

It was not until the publication of Scottish Natural Heritage's *Visual Representation of Windfarms: Good Practice Guidance* (SNH Guidance) in 2006, almost ten years after this technique first appeared, that any explanation was given on the origin of the viewing distance or how it was calculated. However, the controversial wide panoramic format used for both forms of assessment still remained; the only difference was an increase in image height.

The images in environmental statements are now usually presented on A3 pages made up of two or three panoramas with angles of view generally ranging from 73° to 90° including a photomontage, along with a wireframe image which is a computer generated three-dimensional grid showing the landscape profile devoid of any vegetation or structures as shown in Fig. 10 and 11. There is now also a tendency to further increase the size of the panorama on extended A3 pages, which are then folded into the document.

THE BACKGROUND

I first became aware of the use of a specified viewing distance applied to a panoramic image in 1996 in connection with a public inquiry in the Highlands of Scotland when I was still based in the Far East. Although I had been involved in producing computer visualisations for planning submissions for four years by this time, I had never heard of this method of assessment.

From the outset, I found that the technique did not work. It was not possible to focus properly or even hold the image at an exact distance, the images themselves appeared distorted and compressed, the poor quality of the photography made any assessment meaningless, and the images blotted out the real landscape if viewed 'correctly' in the field. In many cases, it was initially difficult to believe I was even viewing the same landscape when the images were assessed from the actual viewpoint. I could not understand why it was necessary to view an image in this way when an appropriate photograph could be simply compared within the real landscape.

In the course of my work in the Far East, I consulted a number of landscape architects on this 'science' as a recognised means of assessment. None had ever heard of it, and all questioned how it was possible to view an image at an exact distance away from the eye. My initial research into the application of a precise viewing science for landscape assessment also drew a blank; however, I felt that the method must have some credibility as its importance was being referred to by a small core of leading landscape architects who were acting as the developers' witnesses at public local inquiries.

The methodology was also misleading experienced photographers. In 1998 during a vacation in Scotland, I was contacted by another community group in the Highlands, who felt that the visualisations submitted in a planning application for a nearby windfarm were deceiving. Because of the costs involved, they could only view the environmental statement in their local council facility and could not test the images on site. However, they instinctively felt that the photomontages somehow looked wrong and underestimated the scale and grandeur of the landscape. One of the group members who was an experienced photographer managed to borrow a copy of the bulky environmental statement. He had a good knowledge of camera lenses and noted that the focal length of the image was stated as 50mm in the technical information, but because the photographs were stitched together forming a seamless image, he naturally assumed he was looking at a 50mm photograph which had been vertically cropped. As I already knew that the 50mm single frame photograph only formed the central part of a much wider panorama, I realised that he was viewing an entirely different and much wider image. A site visit to the nearest viewpoints soon confirmed that he was in fact looking at photomontages which underestimated the landscape scale by around a factor of four.

Over the next four years, the methodology increased in complexity and importance, with no explanation of how it should be used or even calculated. Because of my increasing concerns, I sent a copy of some visualisations based on a focal length of 50mm to the Center for Visual Science in Rochester, USA, who are world-leading authorities in visual perception. To them, the answer was quite straightforward. Even if the images were viewed at the correct distance, they would still under-represent the scale of more distant objects because our brain takes into account distance in the real world to recalibrate their perceived size, a phenomenon known as 'size-constancy'. The problem with viewing photographs involving considerable distances is that our perception of depth is invariably shrunk because we are looking at a flat image with no distance information, so more distant objects will always appear smaller than they do in real life. The use of large foreground objects further diminishes their perceived size.

This simply explained the under-representation I had experienced with my Himalayan photographs many years before. I therefore came to the opinion that the viewing distance was just a convenient red herring; by focusing attention on a complex viewing science which had become an established part of best practice, attention was drawn away from the fact that the images were potentially misleading.

When I relocated to the Highlands of Scotland in 2002, I raised my concerns with Scottish Natural Heritage and The Highland Council. It appeared particularly odd that since the viewing technique had first been brought to my attention in 1996, six years previously, there was no mention of its importance in the Landscape Institute's *Guidelines for Landscape and Visual Impact Assessment, Second Edition* (2002) which had been published.

At the same time, I undertook a series of tests with the help of friends and other members of the public who were prepared to give the matter some considered thought. It was found that single frame A3 and A4 printed images at an increased focal length of 75mm gave them a much more realistic impression. The A3 size was preferred because it could be naturally viewed at a more comfortable arm's length which made focusing easier when comparing the image within the real landscape. In my opinion, the 75mm image still slightly underestimated the vertical scale of the more distant hills, so it was a fair compromise for all parties. Above all, it was an image in a familiar photographic format which could be easily understood by everyone without misinterpretation.

After carrying out observations on how the public viewed displayed images and environmental statements at public exhibitions, the main cause of the problem became clear. Because this viewing methodology was never explained or understood they were in effect, unwittingly viewing a misleading image which considerably reduced the potential visual impact, similar to the way the photographer had been misled by the panoramic images several years before. When questioned afterwards about the viewing distance, most people were completely unaware of it and those who glanced at the technical information considered it to be data for professionals only.

Over the next few years, I investigated many windfarm planning applications which continued to attract claims of visual misrepresentation. In any community there are people who support wind turbine development and those who are just plainly opposed, but the majority of people just want a realistic idea of the potential visual impact. All they want to know is how big and how near the windfarm will be so that they can make a reliable judgement. Because they find that the visualisations presented in environmental statements do not represent what they see, many communities raise funds to commission their own photomontages which are more realistic, though under the Environmental Impact Assessment (EIA) regulations they should have been provided in the first place. In almost all cases their counter-visuals were rejected or not taken seriously by councils or Government reporters/inspectors on the grounds of not conforming to best practice, usually based on the advice of the windfarm developers' consultants who themselves appeared to ignore the relevant guidance.

During this time no evidence could be found to support the viewing methodology for accurate landscape assessment. The only reference located was a small section in the third edition of Sidney Ray's authoritative publication *Applied Photographic Optics* (2002), which provided a theoretical formula based on focal length and magnification. However, he went on to state that 'the correct viewing distance criterion is seldom observed for photographic images, prints being viewed at a convenient distance…'.

Because of growing concerns, in April 2007, I published a paper on the internet entitled *The Visual Issue* (Macdonald 2007) to explain to a general audience the main reasons why panoramic photomontages were potentially misleading. The positive response from all corners of the UK and abroad including Australia, New Zealand and Canada was somewhat overwhelming. Clearly, visualisations were becoming an important issue. As a growing number of problems were becoming evident, I carried out a wide-ranging study into the visualisations presented in environmental statements by using a combination of field tests and computer analysis. Many of these tests were carried out in conjunction with the Planning and Development Service of The Highland Council who were also carrying out their own independent research as a result of ongoing complaints of visual misrepresentation and their own observations on the considerable difference between the original planning visualisations and the emerging built reality. During this period, all research information was shared.

The reason for the lack of any credible scientific evidence of the viewing methodology started to become fully apparent in 2008 when I was commissioned by The Highland Council to undertake a joint investigation into claims of visual misrepresentation made by several community councils in relation to a planning application for a nearby windfarm development. These investigations not only revealed serious problems with the viewing methodology, they also started to reveal a catalogue of technical problems relating to the visualisations themselves. The public claims of visual misrepresentation were found to be fully justified and the applicant was asked to withdraw the application. As a result, a further internal investigation into 26 applications carried out by the Highland Council and Perth and Kinross Council planning officials revealed that none conformed to the guidance cited. This also corresponded with my own investigations into numerous windfarm applications throughout the UK. The problem is threefold: the visualisation methodology is open to misinterpretation; it does not stand up to any technical, practical or scientific scrutiny; and few of the visualisations themselves conform to the recommended guidance. These problems are further compounded by the lack of any prescription or standardisation to enable visualisations to be checked for accuracy. If photography is to be used as a basis for a complex viewing science to support any development proposal, it must be capable of empirical testing. Currently it is not.

A post-construction visual study commissioned by Perth and Kinross Council covering two recently built windfarms in 2009 also confirmed the marked under-representation in the original visualisations. It was observed that even a 70mm focal length did not provide sufficient vertical scale compensation compared to a focal range of 75–80mm. It was also observed that the physical presence of the turbines was considerably reduced in photographic images compared to the three-dimensional experience.

A further problem, particularly for the public, is the availability of suitable visualisations to enable them to make an informed assessment. While it is stressed by windfarm applicants that visualisations are tools to be used in the field, in practice it is made as difficult as possible for the public to do so. Because environmental statements cost several hundred pounds, the only place where the public can view the documents is the local council facility, so the visualisations cannot be taken to site for assessment. Although environmental statements are now generally available on local authority websites or on CD-ROM at a much cheaper price, the visualisations have to be printed at A3 or larger. As most people have an A4 printer for their everyday needs, they can only view the images on a computer screen, where the viewing methodology cannot be applied. Even the viewpoint location maps are invariably so small in scale that they provide no aid in finding the exact viewpoints. The non-technical summaries, short documents which are free to the public, are generally promotional with no visualisations, and in the few cases where they have been included, they are so misleading that any meaningful assessment is impossible.

Photography is both an art and a science. Misleading techniques derived from the art world which diminish the visual impact of the turbines are commonly used, such as the inclusion of large foreground objects and extreme fields of view which create unnatural images with serious distortion. Even selective camera micro-siting, where the camera position can be moved short distances in either direction, is commonly used to deliberately screen turbines with buildings or trees. The quality of photography itself is often unacceptable, with images taken in dull, overcast conditions, which can make the landscape look drab and uninteresting, and the graphic representation and presence of the turbines is often underplayed compared to the lighting conditions or clarity of the background photograph. We tend to appreciate our landscape the most under bright, clear, sunny conditions when the turbines can also appear at their most prominent. It is therefore important that photography is taken in a 'worst case scenario' with the turbine blades facing the camera, as they would be at certain times of the year.

The viewing methodology is simply not understood and there is a widespread level of ignorance of both existing guidance and the basic principles of photography among some professionals, including town planners and even landscape architects. Planning departments also confirm that their scoping requirements are regularly ignored, and the lack of consistency with visuals which are presented in a variety of different sizes and formats is confusing for their committee members and makes any verification for technical accuracy virtually impossible. Most councils therefore accept the assurance of the windfarm applicants that their visuals conform to best practice or the industry standard. As representatives of the public, however, it is the responsibility of any council to ensure that the images presented are accurate, and if complaints of visual misrepresentation are received from the public, they should be properly investigated. Failure to do so may leave the council vulnerable to challenge if the images are later found to have been manipulated to mislead the public, which will become fully apparent once the windfarm is under construction.

To many government inquiry reporters or inspectors, concerns relating to visualisation have generally been regarded as a technical issue. It is my considered opinion that it is much more than this. It is the difference between obtaining planning permission by honest means or by dishonest means using a complex pseudo-science. Viewing distances cannot be accurately applied in practice and are simply being used as a vehicle to technically legitimise misleading visualisations which underestimate visual impact.

In terms of the public role in visual impact assessment, the fundamental problems were clearly identified in the University of Newcastle's *Visual Assessment of Windfarms: Best Practice* under Professor John Benson. The study, which was commissioned by SNH and published in 2002, is explained in further detail in the next chapter. Sadly, Professor Benson died before the next stage of the study could be completed and although the present SNH Guidance (2006) claims to be built upon his initial findings, the independence of academia was compromised by a small group of consultants who largely represented the windfarm industry, and his main recommendations were not further developed or explored.

The general feeling of the public who find the visuals in the environmental statements highly misleading was probably best summed up by a lady crofter from the north of Lewis in the Outer Hebrides of Scotland. In response to the visualisations for the original giant Lewis Windfarm application, she described the turbine representations as 'Nymphs dancing on the moor ... they must take us for fools'.

In an independent publication, *Visualization in Landscape and Environmental Planning*, Professor Benson wrote just prior to his death in March 2004: 'Our main conclusion on the causes of this widespread effect of under-estimation – small photomontages – appears at first sight to be so obvious and simple that it is remarkable that what seems to be the current practice, at least in the UK, is allowed to continue relatively un-remarked, unchallenged and unresolved'. This is still the situation today. In the succeeding chapters of this book it is therefore my intention to remark, challenge and try to resolve some of many issues surrounding current practice, but before the issues involved are investigated, it is first necessary to give an overview of the relevant guidance and advice on the subject.

2

THE FRAMEWORK OF EXISTING GUIDANCE

EUROPEAN EIA DIRECTIVE

 It is important at this stage to briefly review the planning policy framework which relates to the production of environmental impact assessments (EIA), before looking in more detail at the evolution of specific guidance which relates to the presentation of photographs and photomontage visualisations.

In a nutshell, an environmental impact assessment is just an information-gathering exercise carried out by a developer to enable a local planning authority to understand the environmental effects of a development before deciding whether it should go ahead. Of importance in EIA preparation is an emphasis on using the best available sources of objective information and in carrying out a systematic and holistic approach. In theory, the resulting environmental statement (ES) should allow the whole community to properly understand the impact of the proposals. Visualisations will always be a key part of any environmental statement submitted to the planning process.

The legislative framework for EIA which concerns the assessment of effects of certain public and private projects on the environment is set by the European EIA Directive (Council Directive No. 85/337/EEC and amendments 97/11/EC and 2003/35/EC). Because the UK has a number of different development consent regimes, the EIA Directive has been transposed into UK law through a number of statutory instruments which apply in England, Scotland, Wales and Northern Ireland.

In Scotland the regulations state that the EIA Directive's main aim is 'to ensure that the authority giving primary consent (the competent authority) for a particular project makes its decision in the knowledge of any likely significant effects on the environment. The Directive therefore sets out a procedure that must be followed for certain types of project before they can give "development consent". This procedure – known as Environmental Impact Assessment (EIA) – is a means of drawing together, in a systematic way, an assessment of a project's likely significant environmental effects. This helps to ensure that the importance of predicted effects, and the scope for reducing them, are properly understood by the public and the relevant competent authority before it makes its decision.' (www.scotland.gov.uk/library2/doc04/eia-01.htm)

Regulations have recently been further amended to transpose into planning legislation the requirements of Article 3 of the European Council Directive 2003/35 (the Public Participation Directive), which amends the public participation and access to justice provisions of the original EIA Directive. A statutory pre-application community consultation process has also now been introduced into planning legislation across the UK. Meaningful consultation and public consultation can only be undertaken if reliable visual representations of what is proposed are made available as part of the process.

VISUAL IMPACT ASSESSMENT

This book is about providing reliable visualisations for visual impact assessment (VIA), so it is important for the reader to understand what it is. Visual impacts are essentially about change; change in the appearance of the landscape, the effects on people of the change in available views through intrusion or obstruction, and whether important opportunities to enjoy views may be improved or reduced. VIA forms the most important part of any environmental impact assessment because it relates to people, their amenity, the values affecting tourism and the general public.

Visual assessment of transportation environments received attention in the United States from the 1960s and, with the advent of the National Environment Policy Act (NEPA) in 1969, this work was brought together in *Visual Impact Assessment for Highway Projects* (US FHWA, 1988). The basic principles contained in this comprehensive report are still in place today. Visual impact was then defined as follows:

Visual Impact = visual resource change plus viewer response
Change = visual quality before development minus visual quality after development
Visual quality is composed of vividness, intactness and unity
Viewer response = viewer exposure plus viewer sensitivity

Landscape assessment, on the other hand, is concerned with the landscape itself, its components, regional and national distinctiveness and any designations, conservation sites and cultural associations. It includes an assessment of impacts upon the landscape fabric, character and quality. The defining difference between landscape assessment, landscape impact assessment and visual impact assessment is that the first two disciplines are generally the concern of the professional assessor or landscape architect, whereas VIA concerns both the professional assessor and the general public.

In the United States, VIA continues to be a much more thorough process, with a strong emphasis on community involvement. In the UK, however, it has undergone gradual metamorphosis as the landscape profession has adopted a more systematic approach to assessment analysis and the distinction between landscape assessment and VIA has become blurred. VIA is now rarely defined as a single concept and 'impact' is more commonly referred to as 'effect', because the former term was felt to be prejudicial. Landscape and visual matters in environmental statements are now contained in a landscape and visual impact assessment (LVIA), where the lengthy written categorisation, analysis and specialist terminology have become increasingly remote and excluding for the lay audience.

Visual effects are best conveyed visually and not in words, so reliable and accessible visualisation tools are essential. For the professional, such tools are multifunctional and only the start of a clearly defined design and assessment process. Once a proposal reaches the planning process, however, the primary prerequisite of any visual presentation is to communicate the nature of the proposal in such a way that it will be readily understood by the general public and will aid well-informed decision making.

For those interested in VIA and its history, I would recommend a comprehensive reference document on the Macaulay Land Use Research Institute site (www.macaulay.ac.uk). The Task Three VIA Review paper can be accessed by typing 'visual impact' into the search facility on the home page. For an insight into how things are done in the United States, the Macalester College research project site (www.macalester.edu/windvisual/index.html) is also of some interest.

RELEVANT GUIDANCE AND STANDARDS

Guidelines on the Environmental Impacts of Windfarms and Small Scale Hydroelectric Schemes, **Scottish Natural Heritage (2001)**

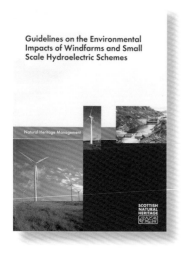

This document was produced in-house by SNH in 2001 and is still cited as a reference in some ES landscape and visual impact assessments (LVIA). It recognises that the communication of information is an important element of the assessment process. On page 17 under 'Presentation of information within a landscape and visual impact assessment' it identifies squarely one of the main issues currently in contention, stating that 'Photographs and photomontages can be useful in illustrating visual character; however, these do tend to underestimate the true visual impact of a windfarm, as the eye tends to focus on certain features and ignore other elements rather than give all constituents equal visual attention. This means that although a 50mm camera lens is commonly accepted to best represent the naked eye, selective focusing by a viewer, particularly to a prominent focus such as a windfarm, may mean a telephoto lens of around 80mm is more truly representative.'

Guidelines for Landscape and Visual Impact Assessment, **Landscape Institute and Institute of Environmental Management & Assessment (2002)**

These guidelines have been the basis of best practice for the production of visualisations for the last decade, and they continue to be the standard work on landscape and visual impact assessment good practice and are widely quoted in environmental statements.

Guidance relating to photomontage visualisation is limited to a few pages in Appendices 8 and 9. Appendix 8, Computer Presentation Techniques, explains in basic but clear language the technical process involved in producing photomontages with computer models. On page 152 it states that 'Photomontages may be commissioned for a variety of purposes, for example as marketing images conveying a general impression of a proposal (where they should be described as for illustrative purposes) or as technically accurate photomontages designed to conform to the rigour of planning applications and public inquiries. The latter require painstaking attention to accuracy and detail. However, as both products may appear graphically similar it is vital that all parties understand the distinction between each, the associated costs and time to prepare and the end use to which they will be applied.'

In Appendix 9 they recommend the use of a 50mm lens and explain that if a wider field of view is required, this can be achieved by a series of overlapping photographs. They go on to state that 'if a practitioner wishes to use an alternative focal length, then a 50mm photograph of the same view should be included for comparison'. During the six years prior to publication of these guidelines, the application of a viewing distance had developed into a highly exacting science which, by 2002, had become an issue at some public inquiries and may well have influenced planning decisions. It therefore seems surprising that no mention is made of viewing distances or their application in this publication.

The guidelines recognise the purpose of the EIA Directive and the need for balance, stating that 'The environmental statement should be an independent objective assessment of environmental impacts, not a best case statement for the development.' General principles of good practice on pages 17 and 18 identify the need for understanding by a lay audience, and Appendix 3, which outlines the Institute's review criteria, states that information should be 'presented so as to be comprehensible to the non-specialist'. It is difficult to fault the spirit of the principles in this guidance; however, they are rarely applied in practice, and with respect to the visualisation element of LVIA, they have been given little regard.

Work is now under way on a revised edition of these guidelines, which are also known as the 'Blue Book'. A consultation among landscape architects has taken place on the proposed structure, which currently includes two sections relating to visualisations, in Chapters 4 and 9. The revision presents an opportunity to look more closely at current visualisation practice and to set standards to meet the needs of all audiences. The revised edition is due for publication in 2012.

Visual Assessment of Windfarms: *Best Practice*, University of Newcastle (2002). Scottish Natural Heritage Commissioned Report F01AA303A

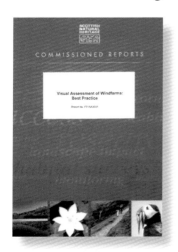

Because of the widespread variation in the way visual assessment was dealt with in EIA documents and criticism of the visualisations, Scottish Natural Heritage (SNH) formed a steering group in 2001 with a view to establishing better and more consistent visual impact assessment in environmental statements.

The University of Newcastle were commissioned to undertake a study under the leadership of Professor John Benson, the Head of their Department of Architecture, Landscape and Planning. The study includes an overview of all the relevant research, policy, guidance and opinion on the subject and a useful and very balanced discussion of the key factors affecting visual effects and their assessment.

The main body of the report is a series of eight case studies of the first generation of Scottish windfarms built between 1995 and 2001 with turbine heights ranging from 54 to 86 metres. It involved visiting 70 viewpoints and comparing the appearance of the windfarm on site with the verbal description and photomontages presented in the environmental statements. The findings confirmed there was a considerable discrepancy between the real-life visibility and the predictions. The conclusions are instructive and call into question many of the photomontage techniques adopted by the windfarm industry.

From the outset the report recognises the contradictions in much of the research and guidance and that one of the factors involved was 'probably the desire of one group of windfarm interests to seek to minimise the political, professional and public perception of the potential visual (and landscape) effects of windfarms, and an opposing desire by another group of interests to maximise these perceptions' (para. 3.2.15). It also recognises the conflict which inevitably occurs between the role of the professional as an expert and their role as an advocate (para. 3.9.9). Professor Benson mentions a number of times the need for more research and post-development audits and how the results and conclusions of useful studies which do exist have failed to penetrate into current practice.

The study was undertaken over a short period of two months, and the time constraints are recognised. However, in terms of visualisations and the wider audience for visual impact assessment, it makes a number of important observations and recommendations, namely:

1. 'Visual Impact Assessment is an integral but distinct part of Landscape and Visual Assessment, and should be distinguished from Landscape Assessment, including Landscape Character Assessment, Landscape Sensitivity and Landscape Significance' (para. 7.2).

2.   '...the related but distinct area of Visual Assessment is as much a matter for people as it is for professionals' (para. 6.3.2).

3.   'If viewpoints are also used as part of any landscape assessment, this should be clearly distinguished from the visual assessment' (para. 7.4).

4.   With regard to camera siting at viewpoints it recommends that 'Precise selection on site should be made to avoid detailed positioning which underestimates the visual effect by the judicious positioning of screening objects' (para. 7.4).

5.   'The limitations of photomontage should be recognised and acknowledged, especially a tendency for photomontage to consistently underestimate the actual appearance of a windfarm in the landscape' (para. 7.5).

6.   'Use of a 50mm lens in a 35mm format is recommended, or equivalent combinations in other formats' (para. 7.5).

7.   'A full image size of A4 or even A3 for a single frame picture, giving an image height of 20cm, is required to give a realistic impression of reality' (para. 7.5).

8.   'A typical comfortable viewing distance for reading A4 pages is 30–40cm, and a typical comfortable viewing distance for larger images at either A4 or A3 held at arm's length is 50–60 cm' (para. 6.1.21). 'A natural viewing distance of 30–50 cm should dictate the technical detail of their [photomontage] production' (para. 7.5).

9.   Because the viewing distances on the panoramic images (17–24cm) would in reality be viewed from greater distances 'a subtle but powerful under-representation of the visual effect is introduced', so what is comfortable for the viewer should 'dictate the technical detail and not vice versa' (paras. 6.1.20 & 21).

The study concluded that 'the increasing development pressures for windfarms require that VIA is approached in a comprehensive, explicit and systematic way and that the inherent complexity, controversy and uncertainty are addressed' (para. 7.10).

Regrettably, Professor Benson died suddenly in March 2004 before the final part of the best practice guidance could be completed, and the University of Newcastle withdrew from any further involvement in the project in September 2004. The study is still referenced in a number of environmental statements. It is available to download from the SNH website and is well worth reading as an authoritative background to this subject.

Visual Representation of Windfarms: Good Practice Guidance, Scottish Natural Heritage (2006)

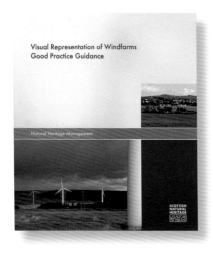

Although this guidance was published in March 2006, it was not made available to the public until March 2007. Instead of appointing another university to maintain impartiality and academic independence, SNH contracted the further development of the visual representation guidance to two members of their original steering group, a landscape architect employed by a firm of landscape consultants who undertook work for the industry and a director of one of the visualisation companies involved in producing the visual techniques which had attracted criticism. The composition of the steering group, which can be found on page 3 of the SNH Guidance, was made up of representatives of SNH itself; consultants who were employed by windfarm applicants; the Scottish Renewables Forum, who represented the interests of the industry; and one representative from the Scottish Society of Directors of Planning, who was the then Planning Director of The Highland Council.

The aims of the Guidance were 'to enable reliable, objective, consistent and accurate visual analysis and visual representation within the EIA process'. It states that it is derived from the research reported within the University of Newcastle's, *Visual Assessment of Windfarms: Best Practice* study; however, 'the sections of this original work concerning visibility maps, viewpoints and visualisations have been updated and refined through a review of current VIA practice, current illustrative methods, consultation with stakeholders and reference to other guidance documents'. In short, the study findings were significantly moderated. In this process the requirements for the public audience became obscured. On page 15 (para. 20) it states, 'This document is not targeted at the general public, given its specialist nature and technical content.'

It is a somewhat lengthy and complex technical document amounting to over 200 pages and is focused entirely on the needs of professional landscape practice. It is noteworthy that on page 18, VIA is redefined as 'the professional and methodical process which is used to assess the impacts of a proposed development on the visual appearance of a landscape and its amenity' (para. 24). In this definition the emphasis is solely on a professional process and appears to exclude the public response which is embedded in the UK planning system.

Despite claiming to have built on the findings of Professor Benson, only his original recommendation of a viewing distance of 300 to 500mm was taken forward. The suggestion that what was comfortable for the viewer 'should dictate the technical detail and not vice versa', was in fact translated vice versa. His observation that a comfortable viewing distance of 500 to 600mm (para. 6.1.21) for larger images at A3 becomes a minimum distance of 300mm, and the minimum image height of 200mm becomes 'over 130mm'. In reality, this enables practitioners to maintain the A3 page format with two images per page and this is now the most common A3 presentation format in the majority of applications.

The recommendation of full-page, single frame images to provide more realism was dropped in favour of the panoramic format as the status quo, the only difference being that the images were now to be slightly bigger. Considerable space is devoted to printing and the production implications of various page size formats with no reference to how large format images would either be used in the field or made available to the public.

This panoramic format is recommended for all aspects of LVIA, giving no recognition of the distinction between landscape assessment and visual impact assessment and the public's role in the latter. The viewing distance is explained in some detail in technical appendices A and B, and on page 73 (para. 126) it states with emphasis that 'not only must the viewing distance be correct, but it must also be set at a comfortable distance'.

Single frame images are not completely excluded. They are included in Figure 37, page 120, with the proviso that they need to show the key characteristics of the visual resource, but all references discourage their use on the basis of inadequate context and the consequent requirement for more than one viewing distance within the environmental statement. Table 17, on page 138, gives a summary of the minimum and preferred requirements. Predictably, it is the minimum requirements which currently provide the basis of visualisation practice.

When I raised concerns on these issues with SNH, I was informed that these compromises had been agreed during two meetings with Professor Benson just prior to his death. However, there is no evidence to clarify the context under which these compromises were made as no minutes were taken or any correspondence exchanged.

This claim would also appear to contradict material produced at a later date. The draft guidance handed over by the University of Newcastle team still listed all of his original recommendations. Professor Benson's chapter in *Visualization in Landscape and Environmental Planning*, published following his death, clearly reiterated his original findings and strengthened his criticism of the existing practice.

The SNH steering group's own spatial data analyst also expressed concerns during the final consultation process that Professor Benson's main recommendation of single frame images seemed to have been 'set aside too lightly, by being lost amongst the other options. It has some very strong points in its favour', namely, it is a familiar format which affects our perception, the field of view is similar to a computer screen and it is the easiest and lowest cost visualisation to produce. He goes on to state that if an objector to a windfarm presented visualisations in a single frame format it would be very hard to reproach them. He also commented that the guidance should take into account that most people will in future view the images on a computer screen. This very important aspect of presentation was never given serious consideration.

Perhaps the key omission lies in the University of Newcastle's handover draft, in which it was recognised that there was 'maybe a need to distinguish between photomontages for context (which may be wide angle and therefore difficult to print for correct viewing distances) and those which are intended to be "realistic" and represent reality/impact which should be printed at the correct size for comfortable viewing'. This would appear to confirm Professor Benson's original observations that single frame images viewed at a comfortable distance gave the best representation of reality.

Overall this guidance represents a lost opportunity to develop a simple and reliable visualisation method to meet the needs of all audiences within the planning system and to establish a fixed verifiable standard. It also contains many unacceptable technical errors. The content predominantly reflects the methodology favoured by some landscape architects, but apparently not used by them. It is readily admitted that photomontages are not generally required by landscape professionals for their assessments but that they may help illustrate the visual effects to the wider audience, yet scant consideration is given to whether the recommended presentation format or viewing technique is suitable for the wider audience.

This guidance forms the current basis of best practice for windfarm visualisations and is cited in most applications. It is available from the Scottish Natural Heritage website although the larger format image examples cannot be downloaded. At the request of the Scottish Government, SNH are now undertaking a major review of this guidance in order to find a more satisfactory solution which will combine the visualisation needs of professional practice and the planning process. It is anticipated that this review will be completed in 2012.

Visualisation Standards for Wind Energy Developments, published by The Highland Council (2010)

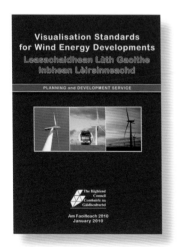

In January 2010, The Highland Council's Planning and Development Service published their comprehensive *Visualisation Standards for Wind Energy Developments*. These standards are a direct consequence of widespread complaints received from the public and community councils regarding misleading visualisations submitted in environmental statements and the discrepancy revealed between the visualisations submitted and the built reality.

The Highland Council first drew up independent visualisation recommendations in conjunction with SNH in 2003. Their *Interim Recommendations for Visual Impact Assessment of Windfarms* (2003) introduced some enlightened recommendations for the use of both 50mm and 70mm focal lengths, minimum viewing distances of 30-50cm with minimum image heights of 20cm indicating an early understanding of the findings of the University of Newcastle Study (2002). (See Appendix 6)

This local authority, which covers a land area larger than Belgium, has extensive experience of windfarm developments dating back to the early 1990s, with over 40 operational and approved schemes and a similar number within the planning process. Both the pressures of development and their experience over the last 15 years has accentuated the need for a reliable standard which can be verified and responds to the growing demand for images which can be clearly and easily understood by the public and planning committee members.

Behind the standards are four years of investigation, empirical testing in the field and research by the council's Planning and Development Service and they currently take precedence over all existing guidance for all windfarm applications within the region. Based on this research, The Highland Council first produced their draft Visualisation Standards in June 2009, with a further revision in November of the same year, which led to an illustrated version of their final Standards in January 2010.

The Standards require the provision of single frame visualisations at 50mm and either 70 or 75mm focal lengths, depending on distance. They also introduce the use of 50mm transparencies and fixed image sizes at ten times (360mm × 240mm) enlargement for an A3 page. The Standards are concise, clearly laid out and also cover requirements relating to viewpoints, overall and detailed map information, photographic and photomontage standards and visualisations for non-technical summaries. The submission of camera metadata is also a fundamental requirement. Panoramic visualisations for professional use are not excluded but some prescription is added to improve technical accuracy and understanding of the necessary viewing methodology.

These Standards are now used as a template for standards being developed by other local authorities. Research in this field is ongoing and the council continues to test visualisation techniques for cumulative impacts, which are a growing concern in parts of the region. A detailed photographic focal length perception study has also been commissioned from the University of Stirling, involving a large sample audience, seven different focal lengths and six viewpoint locations, to inform future policy in this area, and the preliminary results are reported in Appendix 5.

The Standards are expected to be revised and updated as their research continues and in response to feedback from all parties. They can be downloaded from The Highland Council website by searching for Highland Council Visualisation Standards for Wind Energy Developments.

Photography and Photomontage in Landscape and Visual Impact Assessment (Advice Note 01/11), The Landscape Institute (2011)

This Advice Note for landscape professionals was published in March 2011 by the Landscape Institute and superseded Advice Note 01/09. The rationale behind the update is to provide advice on recent technological changes in photography, and a new technical appendix on digital photography in landscape and visual impact assessment work has been added.

Much of the content from the original advice note is reprocessed unchanged, although the language has been altered and even less prescription is suggested. At the outset it recognises the importance of photography and photomontage in the EIA and planning process but with emphasis on their technical nature, recognising that they are only an approximation and proper judgements can only be reached by visiting the viewpoint locations. Under 'Objectives' the overall aim of photography and photomontage is defined as 'to represent the landscape context under consideration and the proposed development, both as accurately as is practical'.

A strong endorsement of the SNH Guidance has been added and, somewhat controversially, a cautionary statement about regulatory authority visualisation requirements. It states that when they 'specify their own photographic and photomontage requirements, the landscape professional should carefully consider whether they are justified, or whether they would under- or over-represent likely effects, in the professional's opinion. Consideration may then be given to adding images to the impact assessment, or omitting them, and explaining the reasons for doing so.'

The General Principles of the previous Advice Note are reworded as 'Criteria for photomontages', which includes that photomontage should 'be based on a replicable, transparent and structured process, so that the accuracy of the representation can be verified, and trust established' and should 'be easily understood, and usable by members of the public and those with a non-technical background'.

The photographic advice is, however, much less convincing, and the reluctance to define a base photographic standard is totally at odds with the stated criterion that the accuracy of representations should be capable of verification. The Institute's solution to the confusion of digital camera technology is to present photographic advice which is almost impenetrable and adds considerably to the existing confusion. The use of conventional 35mm film and a 50mm focal length as a reference standard is described as somewhat outdated. Instead of adopting the digital replacement with a full 36mm × 24mm sensor as the obvious professional base standard, which retains the technical characteristics of the 35mm format and allows verification by others through the analysis of camera metadata, they appear to recommend the use of almost any good-quality digital camera and any focal length, rendering verification almost impossible.

Single frame images are not excluded but it is emphasised that they are unlikely to provide sufficient context. Both wide-angle and zoom lenses are suggested, although the SNH Guidance and most other authorities on the subject do not recommend the use of wide-angle lenses or zoom lenses because of their inherent distortion.

The data incorporated in the technical appendix also appears to be designed to further confound even the well-informed professional by including unnecessary and over-complicated technical information. Any advice which so contradicts its own criteria and presents a jumble of confusing options which defy clear understanding only serves to defeat its purpose. By simply specifying the use of RAW file formats taken on a 35mm digital camera in the first place, all this confusion could easily have been avoided. This document can be downloaded from the Landscape Institute website.

SUMMARY OF RESEARCH PAPERS AND PUBLISHED OPINION

The body of research and opinion in this field is not extensive but there are a number of contributions which are of particular relevance.

Stevenson and Griffiths (1994) undertook a post-construction audit of eight windfarms in England and Wales and noted the importance of image size in representing reality. Sparkes and Kidner (1996) identified that the aims of an environmental statement were to aid the public understanding of likely visual intrusion, which necessitates the use of visual techniques that are easy to understand. Photomontages were described as accurate, quickly understood and not open to misinterpretation. In the same year Roger Cartwright's study for Cumbria County Council of six windfarms (*Visual Impact Study*: *Cumbria Wind Energy Developments*) noted that normal photographs taken with a 50mm lens considerably underestimated the size of the turbines.

All of the above studies related to single frame photography and to turbine heights about half those now being constructed. Professor Benson would reconfirm many of these findings in his *Visual Assessment of Windfarms*: *Best Practice* (University of Newcastle) six years later.

Kay Hawkins and Phil Marsh in their 2001 BWEA paper 'The Camera Never Lies' clearly identified many of the key issues. They recognised that the different audiences may require different presentation formats and that a 70mm focal length may best capture the detail of what the eye sees. The Symonds Report in 2004 went further in a visual audit of the North Hoyle offshore development, finding that the most accurate impression of the perceived view was recorded with a 70 or 80mm focal length and concluded that current best practice guidelines for the preparation of visualisations were misleading.

The 2012 University of Stirling Photographic Focal Length Perception Study, which involved a large sample across six viewpoints, also confirms that a range of 70 to 80mm is overwhelmingly preferred to represent scale and distance in the images when compared with the real scene (See Appendix 5).

The research trail provides clear signposts to the reasons underlying the problems experienced with windfarm visualisations. Yet over almost two decades a range of findings and helpful conclusions have continually failed to influence visualisation practice or to generate further investigation.

Relevant details of all the above papers, studies and many other research contributions are contained in Appendix 6.

OVERALL SUMMARY

The majority of guidance, research and opinion recognises that EIA is a process of prediction and that the information contained in an environmental statement is intended to ensure that before a decision is made the relevant competent authority, statutory consultees and the general public understand the predicted effects.

Current guidance and guidelines have been formulated mainly by landscape architects within the confines of their preferred panoramic methodology, which is designed to provide the context necessary for landscape assessment. There is an emphasis on the technical aspects of creating and presenting panoramic photomontages at the expense of the quite different needs of the wider public audience. Serious questions exist about the scientific robustness of the recommended viewing methodology, which will be examined in this book.

From 1996 onwards, it is possible to trace some recognition of the importance of providing images which are clearly understood by the public through the various research studies. Single frame photomontages were the norm at that time, but as they were replaced by seamless panoramas, the size of the image became an issue as a result of the marked reduction in the size of the single 50mm frame. By 2001, a more representative focal length was being sought to represent more accurately what we actually see.

The accuracy of a photomontage has more than one factor. It should be accurate in terms of the information it contains and how the development is placed in relation to the landscape. It can also be judged in terms of how realistic the image appears to the viewer. Research studies confirm that realistic assessment can be enhanced using camera focal lengths and image sizes, but it is also well understood that photomontage representations will almost always underestimate the true appearance of a windfarm.

While the majority of current guidance in this field is 'practice based', local authority standards offer the most potential for visualisations specifically tailored for the planning system and the public audience. The 2010 Highland Council Standards are based on extensive field testing and have introduced requirements for single frame images in addition to any wide panoramas used by landscape professionals.

A number of conflicts and contradictions are evident both in research and guidance. The Highland Council Standards, the Symonds Report, the DTI/DBERR Seascape and Visual Impact Report and the US National Academy of Science (See Appendix 6) are all at variance with the current SNH Good Practice Guidance and the Landscape Institute on focal lengths, image sizes and the provision of single frame images.

John Benson observed in his conclusion to the University of Newcastle Report, 2002, that 'Given the wealth of research, guidance and experience revealed by this study (even if some of it is contradictory), we are surprised at the general lack of reference to and use of this material in the ESs examined. This apparent failure of research and even practice-based research to penetrate quickly into EIA and VIA practice is an issue that may need to be examined and addressed' (para. 6.5.3).

In this respect little has changed since 2002 and it is an indictment of all those involved that existing studies have been disregarded and so little visual post-construction research has been undertaken. This is particularly remarkable in a period of proliferating applications, increasing wind turbine heights and the number of constructed schemes where the impact is perceived to be greater than the LVIA predictions. Existing opinion, research and the visualisation techniques available to us leave no room for excuses.

20

ABBREVIATIONS

In the succeeding chapters the University of Newcastle's *Visual Assessment of Windfarms: Best Practice* will be referred to as the University of Newcastle Report, 2002. Any reference to the observations and recommendations of Professor John Benson refers to this Report, unless otherwise stated.

The SNH *Visual Representation of Windfarms: Good Practice Guidance* (2006) will be referred to simply as the SNH Guidance (2006).

The Landscape Institute Advice Note 01/11 will be referred to simply as Landscape Institute Advice Note.

3

THE VIEWING DISTANCE METHODOLOGY

THE ARTIST'S VIEW

The SNH Guidance states that the viewing distance is based on the principle of Leonardo's Window and camera obscura, which were techniques developed over centuries by fine artists to obtain correct linear perspective as seen from the artist's actual viewpoint.

Illustration by Mike Taylor

Figs. 13 and 14. The image on the left is an artist's impression of Leonardo da Vinci using his glass window as illustrated in Taylor's New Principles of Linear Perspective *(1719). The image on the right is an artist's impression of Ibn al-Haytham and his camera obscura in Cairo, Egypt.*

During the Italian Renaissance, Leonardo da Vinci wrote: 'Perspective is nothing else than seeing a place or objects behind a plate of glass, quite transparent, on the surface of which the objects behind the glass are to be drawn.' He found, through experimentation, that by drawing a scene onto a sheet of glass from a fixed viewpoint, the image could be traced onto canvas to give the illusion of depth on a flat two-dimensional plane.

Camera obscura also enabled artists to obtain correct perspective by an entirely different method. Using a darkened room or box with a simple pinhole or glass lens in one side, an inverted and reversed image of the real scene could be projected onto the opposite surface and traced. Although first discovered in the early 11th century by al-Haytham (Alhazen), the technique became more commonly associated with Dutch artists from Delft in the 17th century, most notably Vermeer, although it was used by many other artists until the arrival of the photographic plate.

Fig. 15. In Vermeer's painting The Concert, *c.1658–60, camera obscura is thought to have been used to create the complex geometric perspective.*

Fig. 16. This camera obscura from around 1800 enabled the artist to view the image the right way up using a simple lens and an internal mirror.

Both techniques, although quite different, had one thing in common, which achieved the same result. To obtain correct linear perspective, the images could only be captured through a single viewport. In the case of Leonardo's Window it was one eye, and with camera obscura it was a simple pinhole or elementary glass lens. Today, the photographic camera with its single lens automatically captures linear perspective for us.

LINEAR PERSPECTIVE

Fig. 17. The view of the Nave of Brunelleschi's San Lorenzo in Florence demonstrates the rules of perspective.

Whether we realise it or not, we have learned the rules of linear perspective by experience. We instinctively know when we view a painting by a rather poor amateur artist or a drawing by a young child that there is something not right in the picture. It is because the perspective within the image itself is wrong.

Perspective is the art and mathematics of realistically depicting three-dimensional space on a two-dimensional plane, and the study of the projection of objects in this plane is called projective geometry. As Professor Gregory pointed out in his book *Eye and Brain* (1998), 'it is extraordinary that simple geometrical perspective took so long to develop, far longer than fire or the wheel, yet in a way, it has always been present for seeing, as the images in our eyes are perspective projections'.

The principles of perspective drawing were first set out by Florentine architect Filippo Brunelleschi (1377–1446) and refined by the great Renaissance painters of the period such as Masaccio, Leonardo and Raphael.

The rules are summarised as follows:

- The horizon appears as a line.

- Sets of parallel lines meet at a vanishing point.

- Straight lines in space appear as straight lines in the image.

- Lines parallel to the picture plane appear parallel and therefore have no vanishing point.

Because he was also a silversmith, Brunelleschi first of all painted an image of the symmetrical Baptistery in Florence onto a silvered reflective surface as seen from a known fixed viewpoint. To test his theory, he viewed the mirrored image through a peephole drilled at the vanishing point, which was then reflected onto a smaller second mirror. By then adjusting the distance at which this second mirror was held, he was able to view the image and compare its geometrical accuracy in the context of the real scene by lowering the mirror. In effect, he was simply adjusting the focal length of his 'lens' until both images merged together and he had in fact created the world's first virtual reality. Instead of merging the development into a photographic background as we would today, he superimposed his painting of the building into the real scene.

Illustration by Mike Taylor

Fig. 18.

THE VIEWING DISTANCE

In photographic terms, the viewing distance is simply based on an enlargement of the focal length of the camera lens relative to the printed image size. An image taken with a 35mm format camera fitted with a 50mm lens captures a 3 × 2 image on a 36mm × 24mm sensor, so to view correct perspective, the image would have to be viewed at a distance of 50mm. This distance is too small for anyone to focus on, so it is necessary to enlarge the size.

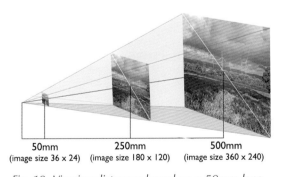

50mm
(image size 36 x 24)　　250mm
(image size 180 x 120)　　500mm
(image size 360 x 240)

Fig. 19. Viewing distances based on a 50mm lens.

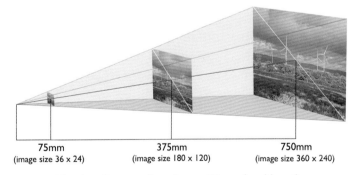

75mm
(image size 36 x 24)　　375mm
(image size 180 x 120)　　750mm
(image size 360 x 240)

Fig. 20. Viewing distances based on a 75mm focal length.

Fig. 19 on the left shows the image sizes relative to distance from the eye using a 50mm focal length. Given the original image size of 36mm × 24mm, it can be seen, for example, that a ten times enlargement will be 360mm × 240mm and have to be viewed with one eye at a distance of 500mm which is a comfortable arm's length and the maximum recommended viewing distance in the SNH Guidance.

The viewing distance can therefore be calculated by dividing the printed image size by the original image size (36 × 24mm) and multiplying by the focal length. In a 3 × 2 single frame, this can be calculated by measuring either the width or the height of the image on the page. If, for example, the printed image size is 360mm × 240mm, the viewing distance can be calculated on the width:

360mm (printed image width) ÷ 36mm (width of original image) × 50mm (the focal length) = 500mm

As the height of the image is 240mm, the viewing distance can also be calculated:

240mm (printed image height) ÷ 24mm (height of original image) × 50mm (the focal length) = 500mm

Based on the same method of calculation, if the focal length is changed from 50mm to 75mm, the viewing distance is increased from 500mm to 750mm, as shown in Fig. 20. Because this distance is beyond arm's length, the wind industry and their consultants now argue that it is not possible to view 75mm images accurately because correct perspective viewing cannot be achieved when the images are hand-held in the field. On the other hand, independent studies have revealed that a 70mm–80mm image viewed at the same comfortable arm's length as the 50mm gives a more realistic representation of distance and scale, so we have a strange conundrum here with two opposing points of view. One is based on our experience of viewing photographs and our ability to compare images, the other is based on the application of perspective geometry and a viewing distance applied with one eye so we can view the photomontage correctly. Who is right? And why is correct perspective so important?

When fine artists applied the rules of linear perspective for their initial sketches, it enabled them to produce an image which gave a realistic illusion of depth. They did not insist that their paintings should only be viewed with one eye from a fixed point at an exact distance, because we have the ability to understand them from a variety of different distances and viewing angles where the perspective looks more or less the same: a phenomenon referred to as 'Zeeman's Paradox'. A drawing containing perspective still appears to be in correct perspective from other viewpoints because we know that a picture is flat and our brain adjusts our understanding of the image according to our viewing position. If there was only one point from which we could view, understand and interpret two-dimensional images, there would only be one seat in a cinema and we all know that this is not the case.

Perspective is therefore not the issue. The important point about a viewing distance is that it projects the correct scale of an object onto the retina; however, we do not see the world around us according to the rules of linear perspective, nor do we see the image projected onto the back of the eye. Our 'seeing' is much more complex because the eye is not only a camera gathering light information, colour and detail; it is also a complex neural organ transmitting millions of electronic messages to the brain, which largely creates the world we see.

Fig. 21.

4

THE EYE AND THE CAMERA

THE HUMAN VISUAL SYSTEM

The human eye is similar to a camera in many ways. The raised cornea acts as a prism gathering and bending light and the hole in the centre of the coloured iris called the pupil restricts the amount of light in a similar way to the aperture diaphragm in a camera lens. This image is then directed onto the lens which projects an inverted and reversed two-dimensional image onto the retina at the back of the eye similar to the image projected onto a camera sensor. Whilst a camera lens achieves sharp focus by using sliding optics, the lens in our eye changes shape according to our focusing distance. As we age, the lens becomes less able to change curvature when we try and focus on close objects, hence the familiar expression that our arms are not long enough when we view printed material or photographic images. Fortunately, however, this can be corrected by wearing contact or spectacle lenses.

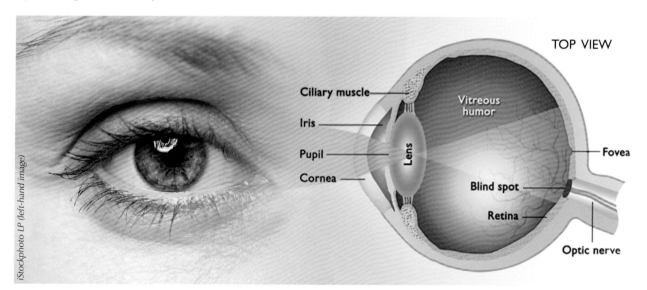

iStockphoto LP (left-hand image)

Fig. 22.

  Although the eye is an optical camera, it is the other parts of the human visual system which enable us to operate in three-dimensional space. The two-dimensional images projected onto the retina of the eye are converted into electrical impulses by millions of highly sensitive photoreceptors which, along with information from the eye muscles, enables the brain to interpret depth. It would be an over-simplification to ignore the mental processes at work when a person sees normally with two eyes. While we obtain many depth cues when viewing with one eye, the fact that we see stereoscopically enables the brain to interpret additional depth information. Closing one eye shuts down the binocular cues.

Under normal viewing conditions, the world appears to us as seen from a virtual eye placed midway between the left and right eye positions. This simple geometrical arrangement has an important consequence because the world appears slightly different from any of these three viewpoints.

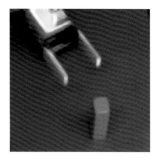 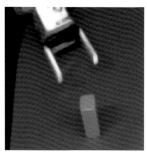 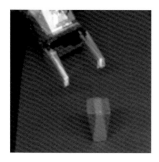 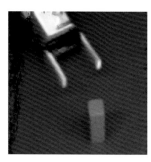

Fig. 23.   Fig. 24.   Fig. 25.   Fig. 26.

Reproduced with the kind permission of the Kiel Cognitive Systems Group and Josef Pauli. http://axon.physik.uni-bremen.de

Fig. 23 is the image as seen by our left eye, Fig. 24 is the image as seen by our right eye. When we view with both eyes our visual system performs a remarkable feat that we do not even notice. It instantly and seamlessly fuses the two images together (Fig. 25) to construct a third image (Fig. 26) as seen by a single eye; the cyclopean eye. In the real world, we rely on a perception of depth to judge distances and the relative scale of objects we see around us. The ability to perceive the world in three dimensions is a trait common to many higher animals. Stereoscopic depth (stereopsis) is the perception of depth which results from the fusion of two slightly different images because of the horizontal separation of our two eyes. This is usually referred to as binocular disparity or retinal disparity.

The fact that retinal disparity is interpreted by the brain as depth was first discovered by Charles Wheatstone in 1838. To prove his ideas, he invented a device called a stereoscope, which showed that a vivid sense of depth can be achieved by viewing two completely flat images of the same scene taken from two slightly different viewpoints. This is the basis of the simple 3D stereoscopic viewers which are still available today and the principle behind the development of 3D film and television. Depth perception also combines several types of depth cues, which can be grouped into three main categories: monocular cues which require the input of just one eye, binocular cues which require input from both eyes, and cues processed by the brain when we see a full field of view with both eyes.

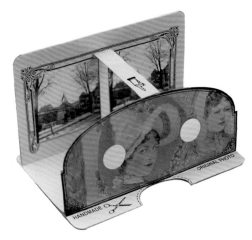

Fig. 27. A replica of a stereoscopic viewer from the late 1800s with side-by-side images taken from two slightly different viewpoints.

Fig. 28. The View-master, developed in the 1930s, was also used by the American military to aid artillery and aircraft recognition in World War II.

HUMAN VISION

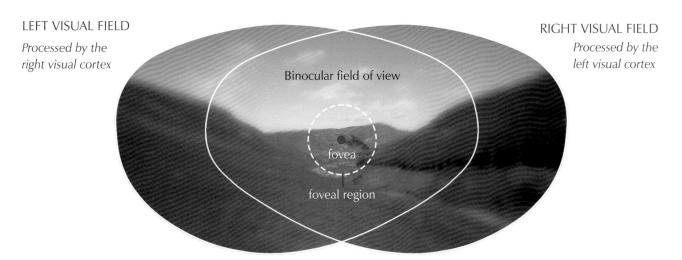

Binocular field of view

fovea

foveal region

Fig 29. *Depending on bone structure, we can see up to 200° horizontally and up to 135°
vertically although we only see objects in detail within the central area of the visual field.
Our binocular field of view is around 120°.*

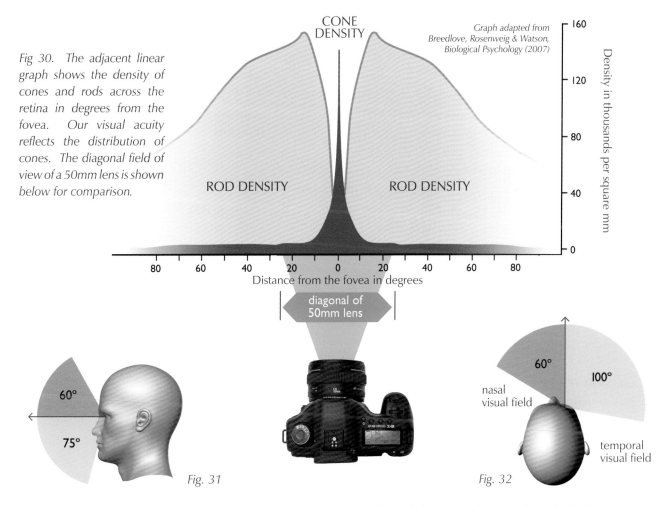

Fig 30. *The adjacent linear
graph shows the density of
cones and rods across the
retina in degrees from the
fovea. Our visual acuity
reflects the distribution of
cones. The diagonal field of
view of a 50mm lens is shown
below for comparison.*

CONE
DENSITY

*Graph adapted from
Breedlove, Rosenweig & Watson,
Biological Psychology (2007)*

Density in thousands per square mm

ROD DENSITY

ROD DENSITY

80   60   40   20   0   20   40   60   80

Distance from the fovea in degrees

diagonal of
50mm lens

60°

75°

Fig. 31

60°

100°

nasal
visual field

temporal
visual field

Fig. 32

*Although the normal field of vision for each eye is about 135° vertically and about 160° horizontally, only the fovea has the
ability to perceive and send sharply focused images to the brain. The foveal field of vision represents a small conical area of
about 1°. Outside a narrow 10° cone, we only see about one-tenth the resolution of our sharpest vision.*

The hemispherical retina at the back of our eyeball which captures the inverted and reversed image projected by the lens is made up of millions of cells called cones and rods. The cones, which mainly form our photopic or central vision, are sensitive to different light frequencies and can therefore detect colour and detail and identify distant objects; the rods, which mainly form our scotopic or peripheral vision, detect only tones of black and white and are less sensitive to detail but very sensitive to light and movement.

The cones which give us our high-resolution vision are concentrated on the central part of the retina within 20° to 25° on either side of our line of sight. The greatest concentrations of cones lie in a circle less than 5mm diameter at the exact centre of the retina known as the macula. In the centre of this region is the 'fovea centralis', our area of highest acuity, which is tightly packed with millions of colour-sensitive cones which rapidly decrease in density as the angular separation from the fovea increases. Although the fovea accounts for only 1° of our vision, most of our seeing is dependent on it rapidly and constantly scanning our environment to build up a detailed 3D image of our surroundings while using finely controlled muscular movements of the eyeball within its socket to focus selectively on things that interest us. Even when we stop momentarily to focus on a particular object, our eyes are still scanning with automatic and extremely rapid movements.

The black and white rods show a different distribution from the colour cones and form our scotopic or dark adapted vision in the remainder of the retina and are mainly concentrated in a ring about 20° away from the fovea. They are much more sensitive to light, which is the reason we look to the side to see dim stars which disappear if we look at them directly. The rods also control the iris of our eye for optimum light, similar to the auto-exposure system on a camera, but they have a much slower adaptation process than the cones, so it can take anything up to an hour in darkness before we achieve optimum night vision. Along with the 120 million rods in the rest of the retina, there are about 6 million more colour-sensitive cones, which combine to send less detailed images to the brain, but from a very wide angle of view, which forms our peripheral vision. However, this region provides much poorer resolution coding of colour, shape and pattern than the fovea but retains highly sensitive motion processing. A much larger portion of the visual cortex is devoted to processing information from the fovea than from the periphery.

Our sight has developed and evolved over millions of years along with our own interpretation of images which we learn from birth. Much of what we see is an interpretation in our brain of the images projected onto the retina and our understanding of them based on experience and memory. People who are blind from birth do not see immediately their sight is restored. It is a long, gradual and difficult learning process. Isolated people who have never seen photographs are confused by them because they have never learned to use the cues of perspective in a two-dimensional image. In his book *Eye and Brain* (1998), Professor Gregory points out that had we not learned the rules of linear perspective, we would simply regard photographs as weird distortions of reality.

Human vision is much more complex than a camera because we have the ability to sharply focus on an object using a very shallow depth of field which provides the brain with precise distance information. Regardless of where you are reading this document, look up from this page for a minute and observe what is in front of you. You will become aware of the fact that we are continually scanning for visual information and detail, building up a three-dimensional image in our brain of the space around us. When we focus on near objects, more distant objects appear out of focus, and when we view more distant objects, near objects appear out of focus. This blur enables us to judge distance and is similar to a very shallow depth of field in photography.

Our ability to rapidly focus on objects at different distances is known as accommodation. If we now explore the other areas to the right or left, we do not rotate our eyeballs to their extremes. We simply move our head to scan for new information while losing visual information and detail on the opposite side. Motion also increases our awareness of objects. In the peripheral area of our visual field, we may not be able to see still objects, but we can detect their movement. In a predatory environment, our lives could depend on that ability. We have evolved to detect motion within the range of animal motion, in other words, we do not see a speeding bullet or a plant growing. This perception of motion is embedded in the visual cortex of our brain, so moving objects such as a windfarm in an otherwise still landscape irresistibly attract the eye and become the point of focus. In an urban environment they are less noticeable because there is more general movement to divert our attention and we are faced with more practical challenges for our own survival such as safely crossing the road.

The distance and visual impact of a windfarm is generally assessed in two ways; stationary, when viewed from a fixed viewpoint such as a dwelling house, but most commonly, by experiencing motion parallax as we pass through the landscape. As the observer moves, nearby objects pass by much more quickly than far-off objects, which appear to be almost stationary by comparison, an effect which is even more pronounced if we travel in a car. Even if we stand still and view a landscape scene, our point of focus relative to our depth of field is also at a much greater distance where we are hardly aware of any foreground objects, so a photograph which has everything in sharp focus from the immediate foreground to the far distance is therefore an unnatural depiction of what we actually see.

Unlike a camera sensor, we do not see the upside down and reversed two-dimensional image projected onto the retina of the eye: we see the world the right way up and the correct way round because the image we actually see has been modified by the brain which facilitates perception of depth. Because we have the ability to assess distance, we scale the size of an object by 'size-constancy scaling'. Put simply, we magnify the size of our retinal image according to its perceived distance in the real world.

An artist who employs strict linear perspective only represents the two-dimensional image projected onto the back of his eye in a similar way to an image captured by a camera sensor. If however, the artist draws what he sees and we now compare his drawing with a photograph from the same viewpoint, we can assess just how far he adopts perspective and how far he draws his scene when scaled for size-constancy. They are likely to be very different, for what we actually see in the real world is affected by constancy scaling, which is our ability to perceive objects maintaining the same size relative to their distance based on our experience. Our retinal image of a person walking away from us gets progressively smaller with distance while a person walking towards us gets progressively larger.

The image in Fig. 33 was taken with a 50mm lens and we can see that the two people are approximately the same height because it is obvious that the person on the left is standing further away than the person on the right. If we compare the size of the same two people standing side by side in the foreground in Fig. 34, the furthest person now appears to be even smaller, yet these are their actual sizes projected onto the retina at the back of the eye. We know by experience that both these people are about the same height because our brain takes into account the distance between them based on the many familiar texture and depth cues within the image itself and enlarges the relative scale of the furthest person.

Fig. 33.

Fig. 34. For a larger image, see Appendix 7 (page 125).

In our retinal image, however, an object halves in size with each doubling of distance of the object, but in the real three-dimensional world it does not appear to shrink anything like as much because the brain compensates for the scale of more distant objects by constancy scaling, where the opposite happens. An object nearly doubles in size with each doubling of distance. (See Emmert's Law on page 32.) Richard Gregory in *Eye and Brain* explains a simple test to illustrate this. Look at your hands, one placed at arm's length, the other half the distance at the elbow. They will appear to be the same size, yet the image of the further hand will only be half the linear size of the nearer. If the nearer hand is now brought to partly overlap the further, they will look very different in size. The overlap defeats constancy scaling, showing what perception would be like without it.

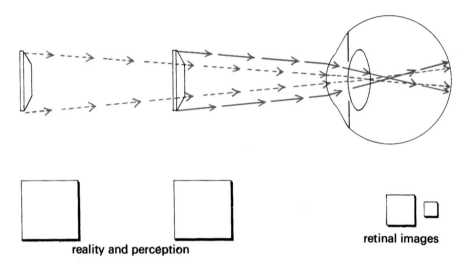

reality and perception   **retinal images**

Fig. 35. Constancy scaling. The image of the object halves in size with each doubling of distance of the object. But it does not appear to shrink anything like as much. The brain compensates for the smaller image with increased distance by size-constancy scaling.

Reproduced from Richard Gregory, Eye and Brain *(5th edition) with the kind permission of Oxford University Press.*

While the image in Fig. 33 shows two people at close distances, our perception of scale becomes more of a problem in a photograph of a landscape involving considerable distances, due to depth compression where everything is in sharp focus and may lack reliable scaling or perspective cues. Because we rely on our perception of actual distance to correctly scale an object, when we view a landscape, our eyes accommodate rapidly over a wide range of distances where our binocular lines of sight become parallel at distances beyond 10 metres. Our visual system uses 'blur' and shallow depth of field to gauge distance, but when we look at a photograph of the same view, our eyes converge and are focused at the distance we are holding the picture. Our eyes tell our brain that the page is, for example, 300mm from the eye, regardless of what illusion of depth is depicted in the image. This process is 'informationally encapsulated' or embedded in our visual system and cannot be overridden by the intellect.

A site test will confirm this. If a 50mm transparency of the landscape involving some distant hills is created where the real distance cues are clearly visible, it will fit the real landscape when viewed with one eye, as shown in Fig. 36. If the transparency is now viewed with both eyes, the image is seriously out of alignment because a viewing distance based on the principle of Leonardo's Window no longer applies. Now view a printed image at the same distance so it can be directly compared to the wider landscape as illustrated in Fig. 37. Whether it is viewed with one eye or two eyes, the vertical scale of the more distant hills is compressed, making them appear to be further away than they are in reality. It is for this reason that foreground objects should be excluded from windfarm visualisations because they will appear disproportionately larger than the more distant turbines.

Fig. 36.

Fig. 37.

In my dealings with communities over many years, I have also come across some interesting ideas on how to create more suitable images for visual impact assessment. For example, a retired geologist found that stretching the vertical scale of a 50mm image gave him a more realistic impression; a forester increased the focal length of his zoom lens from 50mm to around 70–80mm to achieve what he saw. Even in a panorama of his photographs joined together, the vertical scale of the landscape was noticeably different. Both people had an interesting and very valid point, because without knowing it, they were simply applying size-constancy scaling and recognising the difference between what we see in a two-dimensional photograph and what we see in the real world.

Perhaps the best example of the application of size-constancy scaling can be found in Alfred Wainwright's Pictorial Guides to the Lakeland Fells published during the 1950s and 1960s. He had the remarkable ability to represent what we see. Using photographs taken on a Box Brownie camera as a reference for many of his sketches, he increased the vertical scale of the landscape as the distance increased. In an article in the *Westmorland Gazette* in 1966, he said that 'it is necessary only to remember that the ordinary camera tends to depress verticals and extend distances and correct for this imperfection'. The problem could not be explained more eloquently.

Frances Lincoln Ltd. 2004 © The Estate of A. Wainwright.

Fig. 38. 'Little Langdale' by A. Wainwright from A Third Lakeland Sketchbook

EMMERT'S LAW

First, obtain a good clear after-image, by looking steadily at a bright light, or preferably a photographic flash. Then look at a distant screen or wall. The after-image will appear to lie on the screen. Then look at a nearer screen. The after-image will now appear correspondingly nearer – and will look smaller. With a hand-held screen (such as a book) moved backwards and forwards, the after-images will expand as the hand moves away and shrink as it approaches the eyes, though of course the retinal after-image remains of constant size. We therefore see the brain's scaling changing as the distance of the screen changes. The after-image is seen to (nearly) double in size with each doubling in distance of the screen. This is Emmert's law. It is typical of the psychological projection of retinal images into external space.

Reproduced from Richard Gregory, Eye and Brain *(5th edition) with the kind permission of Oxford University Press.*

CAMERA FORMATS

If we aspire to represent what the human visual system sees in terms of scale, photography is the only practical means of achieving this. It is not a perfect technology but it is a practical one which can be reliably adapted to match our understanding of how and what we see. The format recommended by the windfarm industry and all guidance is the 35mm camera format fitted with a 50mm standard or 'normal' lens.

The first prototype camera was developed in Germany by Oskar Barnack in 1914 to make use of cheap film which had been developed for the burgeoning movie industry. By turning the 35mm filmstrip sideways, he found that the frame size could be doubled to 36mm × 24mm, which is still the standard image size today. The first commercially available models did not appear until 1925 with the production of the Leica 1, a robust, well-engineered compact camera with a separate lens and viewfinder, which established the 35mm format as the format of choice. The first 35mm single lens reflex (SLR) cameras, which enabled the photographer to view the actual scene through the lens using an eye-level viewfinder, were developed in the late 1930s in Germany, but they still required the use of a separate light meter to calculate the correct exposure.

By the 1960s and 1970s, advances in optical engineering and electronics enabled manufacturers, mainly of Japanese origin, to produce cameras with interchangeable lenses and built-in light-metering systems incorporated into one compact unit. Among the leaders at the time were Pentax, Olympus, Nikon and Canon, names which are today synonymous with high-quality cameras. Originally, they were usually supplied with a 50mm standard lens, although today, zoom lenses have become the norm. The image quality, however, was restricted by the resolution of the film itself and many photographers upgraded to larger format cameras with bigger frame sizes to achieve more professional results. Today, traditional film has been totally replaced by electronic sensors, revolutionising photography, and the widespread use of conveniently-sized digital cameras and camera phones has made good quality digital imaging available to almost everyone at a reasonable cost.

Fig. 39. The Canon AE-1 (1976).

The digital age has brought about huge advances in photography. Image resolutions have been greatly increased compared to traditional film and many professional photographers have now returned to the convenience of the 35mm format. By simply replacing the 36mm × 24mm film frame with an electronic sensor of the same size, the characteristics of the 35mm format are retained. There is no difference in format size between conventional and digital photography: a 50mm fixed lens, for example, will produce a technically correct 50mm image. One of its greatest advantages is the fact that the camera information is recorded with each frame, so where images require verification, this can be easily established from the camera metadata, which can be read in a variety of different graphics software.

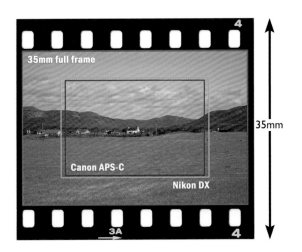

Fig. 40. This graphic shows the size of different sensors in relation to the 35mm frame size of 36mm × 24mm.

The digital age has also brought about a lot of confusion. Today, there are basically two types of camera on the consumer market: small compact zooms with retractable lenses, which are very convenient and easy to carry, and SLR cameras, generally fitted with zoom lenses. The problem is that the term SLR, which we have traditionally associated with 35mm film, need not necessarily mean a 36mm × 24mm image size.

To keep down production costs, many companies produce consumer SLR cameras with electronic sensors smaller than conventional 35mm film, such as the APS-C sensor which retains the same 3 × 2 image proportion. Technically, they are not 35mm cameras and require a multiplication or enlargement factor to give the 35mm equivalent. A 28mm wide-angle lens on a camera which has a conversion factor of 1.5X will give a 35mm equivalent of 42mm. Similarly, a 35mm lens will give an equivalent of 52.5mm. As can be seen in the above graphic, the sensor just crops the image taken by a lens with a wider field of view. Apart from the problems of distortion associated with both zoom and wide-angle lenses, it is not possible to produce a technically correct 50mm image, which is the minimum requirement of the SNH Guidance, so the use of reduced-size sensor cameras should not be considered for professional landscape photography.

In most photographic situations sensor sizes and focal lengths are not an important issue, but in the case of windfarm visualisations, the technical requirements are much more demanding as the photomontages are often referred to as technical drawings. Unlike building developments, where plans and elevations can also be assessed, the visualisation of windfarm developments are the *only* means by which the public can make an informed judgement. In an architect's drawing, floors of equal height are not reduced in dimension as the building gets taller so; in a similar way, the vertical scale of the landscape in a photomontage should replicate what we actually see. We are therefore not simply taking a photograph. Unlike viewing a two-dimensional technical drawing, we have to take into account an additional dimension: the third dimension of distance. Within the limitations of photography, we have to portray a realistic impression of the likely scale and impact of the development when seen from the actual viewpoint.

Careful consideration therefore has to be given to camera lenses so that the images are easily understood and also capable of independent verification by the planning authority. The Landscape Institute Advice Note states that the visualisation representations should be 'based on a replicable, transparent and structured process, so that the accuracy of the representation can be verified, and trust established' and that the visualisations should be 'easily understood, and usable by members of the public and those with a non-technical background'. Standardisation of camera format and sensor size is essential if the images are to meet these criteria and be verified for accuracy.

CAMERA LENSES

With the development of point and shoot cameras, little consideration is given to camera lenses and focal lengths. The scene is simply framed to the desired field of view depending on the subject matter. While we may zoom in to capture a close-up of a person's face, when photographing a landscape there is a temptation to zoom out to the widest setting to capture as much of the scene as possible, only to find that the image of the imposing mountain in the far distance appears to be the size of a small hillock. To understand why this problem occurs, it is necessary to give an overview of the characteristics of camera lenses.

All lenses, regardless of camera format, have three fields of view: horizontal, vertical and diagonal, which vary proportionally according to the focal length. The SNH Guidance (2006) states that the horizontal field of view is the most important and that the diagonal field of view is of little practical use. However, such a statement is fundamentally wrong because all prime or fixed lenses are generally specified by their diagonal field of view and there is a straightforward reason for this.

The focal length of any lens is defined by its diagonal field of view or picture angle because it is the diameter of the circle projected onto the electronic sensor. The sensor lies behind the camera shutter at the back of a rectangular chamber containing the moving mirror which directs the real image to the viewfinder. As the 35mm format has a sensor size of 36mm × 24mm and therefore a fixed aspect ratio of 3 × 2, the diagonal not only dictates the focal length, it also dictates both the horizontal and vertical fields of view.

The table below shows the relative fields of view of different fixed lenses. Different focal lengths play an important part in our perception of scale, particularly in landscapes where considerable distances are involved.

Fig. 41. The diagonal field of view is the diameter of the circle projected onto the camera sensor excluding corner vignetting.

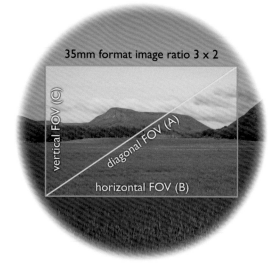

Fig. 42. The diagonal field of view dictates the horizontal and vertical fields of view depending on the aspect ratio of the camera format.

FIELDS OF VIEW OF FIXED LENSES: 35mm FORMAT

Lens focal length	Diagonal (A)	Horizontal (B)	Vertical (C)
24mm	84.0°	73.7°	53.1°
28mm	75.4°	65.5°	46.4°
35mm	63.5°	54.4°	37.8°
50mm	46.8°	39.6°	27.0°
85mm	28.5°	23.9°	16.0°

The above angles of view should be taken as a guide only.

FOCAL LENGTH

The focal length of any camera lens is the distance between the optical centre of the lens and the film or sensor at the back of the camera when the focus is set to infinity. The shorter the distance, the wider the field of view, which makes distant objects look further away, and the longer the distance, the narrower the field of view, making distant objects look nearer.

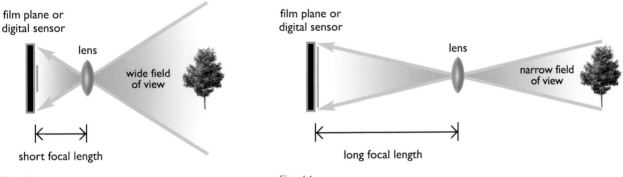

Fig. 43.   *Fig. 44.*

In any camera format, a 'normal' lens has a focal length roughly equivalent to the diagonal measurement of the sensor frame, which gives a natural, unforced perspective similar to what is seen with the human eye. For the 35mm format, which has a diagonal dimension of approximately 43mm, the nearest prime lens is the 50mm normal lens. Camera lenses with a shorter focal length are defined as wide-angle lenses and those with longer focal lengths are defined as telephoto lenses. Different focal lengths also give us different perceptions of distance. If we look at the person in the two images below, our impression of distance is different, yet the photographs were taken from the same viewpoint.

Fig. 45.   *Fig. 46.*

The person in the image on the left looks as if she is only a few feet away because it was taken with a telephoto lens, but in the right-hand image the person appears to be several metres away because the image was taken with a wide-angle lens. Our perception of distance is achieved by cropping the image, retaining the 3 × 2 ratio of the camera format. As the focal length increases, the fields of view become narrower. Providing photographs share the same central axis, images of different focal lengths will fit into each other when adjusted in lens correction software. This is best illustrated in the figure on the next page, which shows the relative fields of view of different prime or fixed lenses.

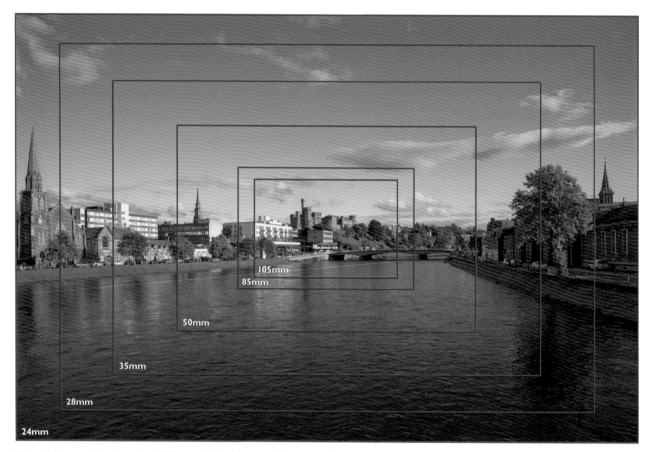

Fig. 47. Image showing the different fields of view of prime or fixed lenses taken on a 35mm format camera.

Images of different focal lengths, however, do not reduce in size relative to the printed page as shown above and our perception of varying distances only becomes fully apparent if all the images are printed at the same height. In the images below, you can now see that our perception of distance varies according to the focal length of the lens. The more the image is cropped, the nearer the castle appears.

24mm wide-angle lens   *50mm normal lens*   *105mm telephoto lens*

Fig. 48.

In a landscape, the focal length has to be carefully considered, as it can have a considerable effect on our perception of the scale of more distant objects. The image on the next page shows the difference in vertical scale using focal lengths of 24mm, 50mm and 75mm. The hill in the distance appears to get nearer as the fields of view are decreased. Which one is most representative of what we would see in terms of perceived distance and scale if we stood at the actual viewpoint? If we increase the angle of view to accommodate the wider landscape, we do so at the expense of our perception of distance, and if we decrease the angle of view to get a more realistic impression of distance, we do it at the expense of the wider landscape.

Fig. 49. A 50mm planar panoramic image with a horizontal field of view of 73°, which is the horizontal field of view of a 24mm wide-angle lens.

Fig. 50. A 50mm single frame image increases the vertical scale of the landscape but still underestimates our perception of distance.

Fig. 51. A 75mm single frame image increases the vertical scale of the landscape and is similar to what we actually see.

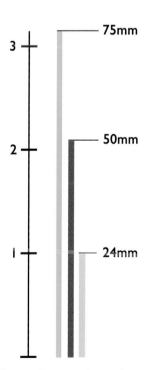

Fig. 52. Comparative scales.

37

THE 50mm LENS AND THE HUMAN EYE

The 50mm lens, first developed by Leica in 1925, was a compromise between diagonal dimension and lens technology at the time, which could produce a sharp photographic image. The slightly longer 50mm focal length was able to achieve optimum sharpness while minimising distortion in the vertical and horizontal directions. It does, however have a similarity to the human eye. Its diagonal field of view of 46° is similar to the field of view of the colour-sensitive cones which give us our high-resolution vision. It is interesting to note that in observations carried out in the field with The Highland Council, the diagonal field of view of a 50mm lens more than adequately represented what was consciously seen in detail by both eyes when selectively focused on a point of interest in a landscape.

However, there is a contradiction here. While a 50mm lens is a fair representation of what we see within our area of detailed vision, it does diminish the scale of more distant objects in a printed image containing linear perspective where the scale of an object diminishes as the distance increases. Because the scale of an object enlarges as our perception of distance increases, a 50mm photograph will always under-represent the scale of wind turbines which are much further away, and the greater the distance, the greater the problem. A more representative sense of distance can only be achieved by increasing the focal length of the camera lens until the vertical scale of the landscape matches what we see, similar to what Brunelleschi did with his mirror six hundred years ago. The diminishing problems arising from linear perspective still arise where long distances are involved, but the effect can be reduced by decreasing the field of view, as illustrated on the previous page.

THE TELEPHOTO LENS AND THE HUMAN EYE

There is an interesting experiment that anyone with access to a professional 35mm format camera fitted with a full-size sensor and a 100% viewfinder can carry out in the real landscape. Using a zoom lens, preferably in the range of 24mm to 105mm, hold the camera sideways to the left eye while still looking at the real landscape with the right eye. This is best achieved by relaxing both eyes and drifting the camera viewfinder over the left eye until you see both images. Adjust the focal length of the camera lens until both images match. Gently rotate your hips to 'follow' the landscape to recheck that the images align with each other. Now check the focal length on the camera lens.

I have conducted many tests over the years using this technique with different 35mm cameras and found the results to be very consistent at around a focal length of 70mm to 75mm. For printed images, however, this focal range appears to be much wider. In 2011, The Highland Council commissioned the University of Stirling to undertake a focal length perception study. The object was to investigate the effect of focal length on public perceptions of scale and distance in landscape photographs. Although the detailed document was not available at the time this book went to publication, a preliminary report was released jointly by the University and The Highland Council. (See next page and Appendix 5.) As an alternative to viewing through the camera lens, if the camera has a liveview facility where the image can also be viewed on a laptop screen, use the zoom lens to vary the focal length in landscape orientation until it appears to match the actual landscape.

Fig. 53.

The preliminary University of Stirling report was prepared after interviewing 362 respondents, but I understand that the results from the completed study involving over 500 participants has revealed similar results. It was found that there is a much wider variation in the relationship between focal length and perceived scale and distance. The preference range was from 70mm to 90mm with a mean average over all distances of 79mm. The findings also suggest that the use of a 50mm single frame image for the visualisation of wind energy developments is inappropriate for realism and the study supports the use of greater focal lengths as stipulated in The Highland Council's *Visualisation Standards for Wind Energy Developments* (2010).

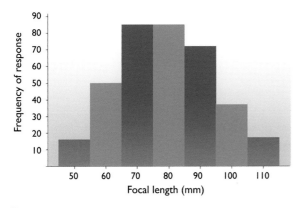

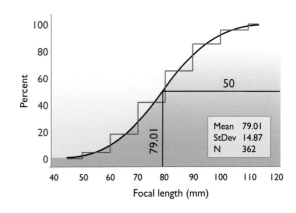

Fig. 54

Reproduced with kind permission of the University of Stirling and The Highland Council.

So why is the camera lens test more consistent in the choice of focal range than the printed images? From my experience, I think there is a simple answer. In the experiment when we view through the lens with the left eye, we are viewing the flat image reflected by the camera mirror onto the translucent focusing screen but with the right eye, we are still able to assess distances within the real landscape. We have a fusion of two images; the two-dimensional image in the viewfinder and a three-dimensional image of the real world which has been adjusted for size-constancy. By now adjusting the focal length of the camera lens, we can accurately match the vertical scale of the real landscape.

It becomes a bit more problematic when we view a photograph normally with both eyes because the distance cues which enabled us to calibrate depth have been removed and we are looking at an image where everything lies on the same focal plane. Our perception of depth is invariably shrunk, which may account for the much wider variation in the choice of focal length.

Fig. 55. 50mm focal length.

Fig. 56. 75mm focal length.

5

DISPLAY AND VIEWING CONVENTIONS

VIEWING IMAGES

The observation and recommendation by the University of Newcastle Report (2002) that what is comfortable for the viewer should 'dictate the technical detail and not vice versa' was a very important point. The flatness and upright orientation of an image are conventional: we know that a photographic image is flat because drawing paper and photographic paper are flat. These are the display conventions contained within perspective geometry and they are expertly described in Bruce MacEvoy's 'Handprint' website (www.handprint.com/HP/WCL/perspect1.html).

We learn this from a very early age. A young child looking at a picture book learns to understand that the images are flat and very different from the three-dimensional world which surrounds them. Like the process of seeing, it is a gradual process where we learn to view pictures at a comfortable distance. Most printing formats therefore assume fairly close viewing distances: book illustrations are normally viewed from a comfortable reading distance and even wall posters are typically reduced for display in smaller domestic spaces.

In a similar way, we instinctively apply the same viewing convention to much larger images, for example, in an art gallery. If we observe the public viewing Rembrandt's colossal painting (4.4m × 3.6m) of the *Night Watch* in the Rijksmuseum in Amsterdam, while they may view the image at closer range to examine the Master's technique and detail, they stand at a considerable distance to comfortably view and appreciate the whole scene. These conventions are so powerful, and so basic to our visual experience, that we even apply them to abstract art.

Ideally, if the viewer is standing squarely in front of the image with head erect and the image is hung at eye height and parallel to the floor, then our display and viewing conventions can be summarised as:

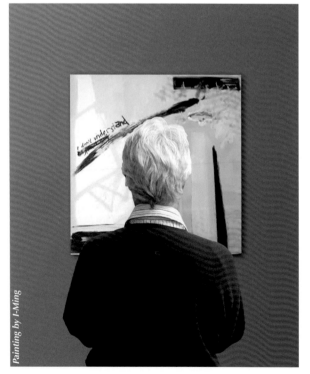

Painting by I-Ming

Fig. 57. Our natural viewing conventions are so powerful that we even apply them to abstract art.

- at eye level

- facing the image

- comfortably far away.

For a hand-held photograph, the larger the image, the further away we hold it. A small-size A5 print will be held quite close to the eyes at a distance where we can see the image in clear focus, whereas a much larger A3 print will generally be held at a more comfortable arm's length.

Fig. 58. When we view a picture, the larger the size, the further away we hold the image to view it comfortably.

Dr Leslie Stroebel, former Professor of Photography at the Rochester Institute of Technology, New York, in his paper *Perspective: Image, Camera, Distance and Objects* also confirms that we tend to view photographs from a distance about equal to the diagonal of the page. He also observed that it is fortunate that people do tend to view photographs from standardised distances based on their size, rather than adjusting the viewing distance between the image and the eye to make perspective appear normal, which is the SNH approach. This also roughly equates to the art-world rule of thumb that the viewing distance should be one and a quarter times the maximum dimension of the image. The imposition of a geometrically exact viewing distance which can only be applied with one eye at an unnaturally close distance is therefore alien to most people, who naturally view a picture at a distance relative to its size with both eyes: a predisposition which is difficult to override. This conflict between our natural viewing convention and an unnatural viewing distance is the main problem leading to the under-representation of visual impact in environmental statements.

THE SHRINKING LANDSCAPE

When we view a photomontage in a landscape, we simply hold the image at a comfortable distance and compare it to the real world in front of us. An image will therefore look too far away, too near or just about right, based on our experience and understanding of two-dimensional photographs. When we stand at a particular viewpoint, does the landscape look:

This near?   *This near?*   *Or this near?*

Fig. 59.

In Chapter 4, it is explained that a printed 50mm photographic image will always under-represent our perception of the scale of a more distant object because we are looking at a flat image devoid of any depth information. However, the main cause of underestimation with the images presented in environmental statements is due to the presentation format and how the images are viewed.

In terms of perception, the problems can be summarised as follows:

1. The landscape is distorted and looks much further away than it does in reality.

2. The vertical scale of the landscape appears to be compressed.

3. The presentation format is unfamiliar.

The best way to illustrate the point is to compare the two visuals below, which would be printed at the same width on an A3 page. Both were taken with the same 50mm lens, using the same camera format, and from the same viewpoint. They should be identical, yet our perception of distance is very different. The upper image looks much further away than the lower image. Why? It is due to three reasons: The use of a panoramic format, a viewing methodology which is not understood or clearly explained, and our natural viewing convention.

Fig. 60.

Fig. 61.

Below is a typical visualisation page from an environmental statement with a field of view of 73°, although images with reduced heights and fields of view of up to 90° are still not uncommon. At the bottom of each page, there is a small box which contains technical information. Among the small print, apart from the focal length of the camera lens, a viewing distance is specified which means nothing to members of the public. The majority of people do not even notice it and if they do, they rightly assume that it is technical data for professionals. They do not use a viewing distance to view images in a magazine or to watch television, so it is of no significance to them, nor do they realise its importance.

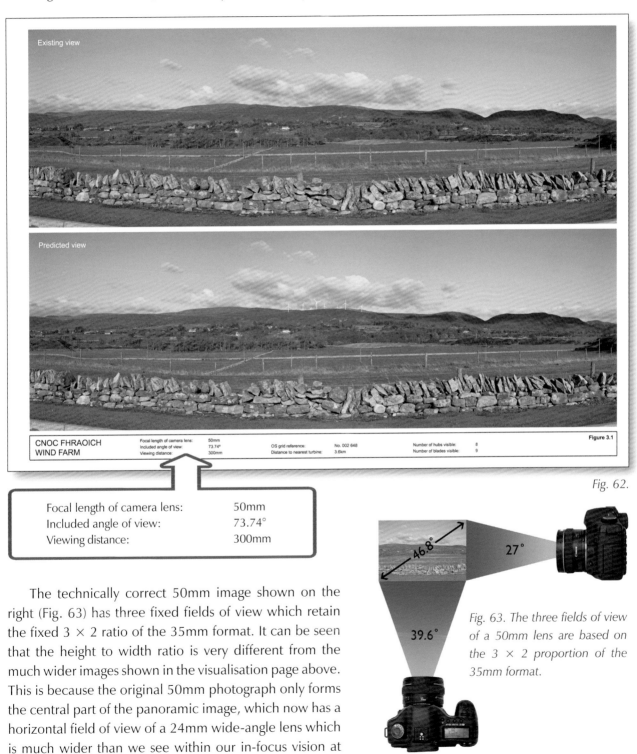

Existing view

Predicted view

CNOC FHRAOICH WIND FARM	Focal length of camera lens:	50mm	OS grid reference:	No. 002 648	Number of hubs visible:	8
	Included angle of view:	73.74°	Distance to nearest turbine:	3.6km	Number of blades visible:	9
	Viewing distance:	300mm				

Figure 3.1

Fig. 62.

Focal length of camera lens:	50mm
Included angle of view:	73.74°
Viewing distance:	300mm

The technically correct 50mm image shown on the right (Fig. 63) has three fixed fields of view which retain the fixed 3 × 2 ratio of the 35mm format. It can be seen that the height to width ratio is very different from the much wider images shown in the visualisation page above. This is because the original 50mm photograph only forms the central part of the panoramic image, which now has a horizontal field of view of a 24mm wide-angle lens which is much wider than we see within our in-focus vision at any given time.

46.8°

27°

39.6°

Fig. 63. The three fields of view of a 50mm lens are based on the 3 × 2 proportion of the 35mm format.

Because there is no information anywhere on the page to explain how the images should be viewed and only viewed, a powerful underestimation occurs which makes the landscape look further from the actual viewpoint than it does in reality, which in turn reduces the visual impact of the turbines. Why this happens is explained in the graphics below. Even if the public were to view the images at the correct viewing distance, two vital pieces of information required for accurate viewing are usually omitted. To view correct perspective, the images must only be viewed with one eye, curved in an exact arc equal to the field of view with our line of sight fixed on the centre of the panorama. The photograph if correctly viewed would completely screen the real landscape.

The next question is how do we know that the single frame image which forms the central part of the panorama is a 50mm photograph, the overall angle of view is 73° and the correct viewing distance is 300mm? The simple answer to that is we don't know because under the current methodology there is no way this information can be verified. The criteria for photomontages in the Landscape Institute Advice Note state that the images should be capable of verification, while the very same Advice Note permits the use of almost any camera, any lens and any focal length, making verification almost impossible. Not only would the planning officer have to look up the camera's detailed technical specification for each application to establish the conversion factor for the 35mm format equivalent, he is also faced with a highly complex best practice which is not universally understood. Planning officers therefore tend to accept the word of the windfarm developer that the images accord with all the relevant guidance, because they have no practical way to check the accuracy of the images.

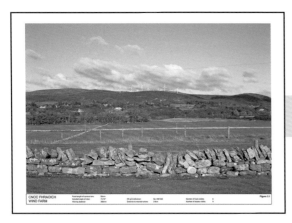

The technically correct 50mm image which should be shown full-frame ...

Fig. 64.

is reduced in size on the A3 page ...

Fig. 65.

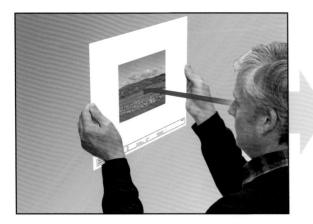

and should now be viewed and only viewed at an exact fixed distance for correct perspective ...

Fig. 66.

images are added to increase the horizontal field of view for wider landscape context ...

Fig. 67.

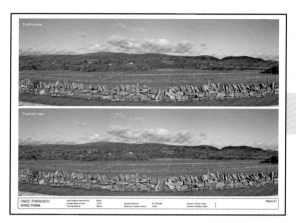

the vertical height of the image is cropped and other images are now added to make up a full A3 page ...

Fig. 68.

which is naturally viewed at a more comfortable distance, which is greater than the specified viewing distance ...

Fig. 69.

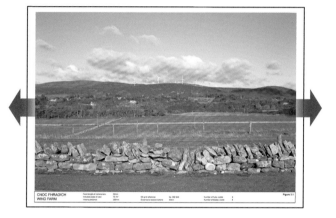

so instead of viewing the original 50mm image with additional peripheral vision of one eye ...

Fig. 70.

the viewer is looking at the much wider image, which makes the landscape look distorted and further away.

Fig. 71.

THE DIMINISHING IMAGE

As explained on page 41, the imposition of a geometrically exact viewing distance which can only be applied with one eye is simply alien to most people, who naturally view a picture at a comfortable distance with both eyes. When we view a photograph, the three angles of view which technically define the focal length of the camera lens remain the same regardless of the distance at which we may comfortably hold the image relative to the size of the print.

By contrast, the 'correct' viewing of a panoramic image is achieved by viewing the image at an exact distance with one eye. Most images presented in A3 environmental statements require a viewing distance of anything up to 200–250mm shorter than our natural viewing distance relative to the page size. Because the image projected by a 50mm lens has a diameter of 46.8° (See Fig. 63), a 3 × 2 image within the circle has a horizontal field of view of 39.6°. The further away we hold the image, the wider the field of view of the 39.6° rectangular cone becomes, introducing the 'subtle but powerful under-representation of the visual effect' identified by Professor Benson in 2002. The extent of the under-representation is illustrated on the next page.

Images based on a 50mm vertical field of view and an image height of 144mm.

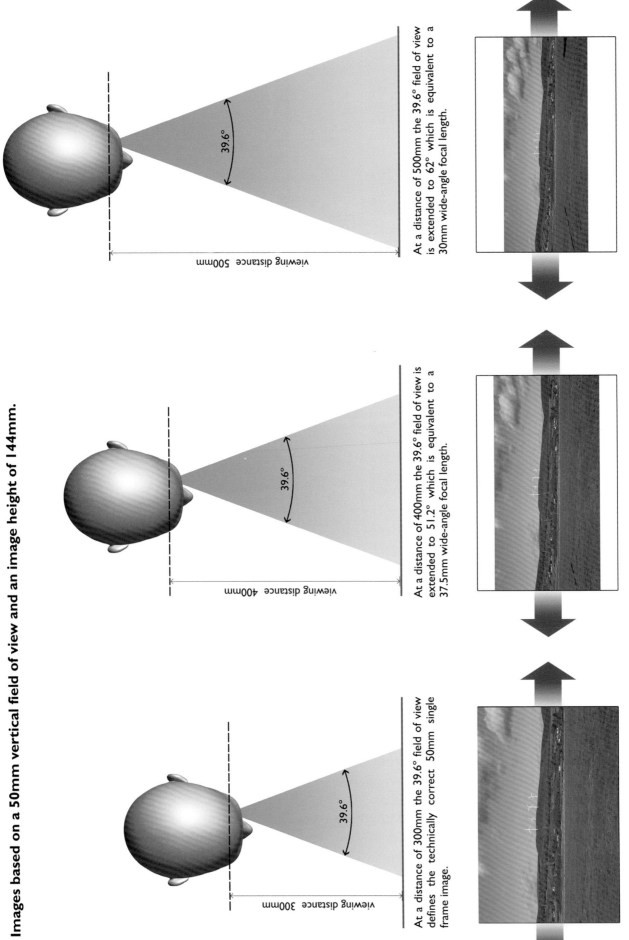

viewing distance 300mm

At a distance of 300mm the 39.6° field of view defines the technically correct 50mm single frame image.

viewing distance 400mm

At a distance of 400mm the 39.6° field of view is extended to 51.2° which is equivalent to a 37.5mm wide-angle focal length.

viewing distance 500mm

At a distance of 500mm the 39.6° field of view is extended to 62° which is equivalent to a 30mm wide-angle focal length.

Fig. 72.

Fig. 73. If the panoramic visualisation is viewed at a greater distance, the 39.6° horizontal cone is now wider than the image page.

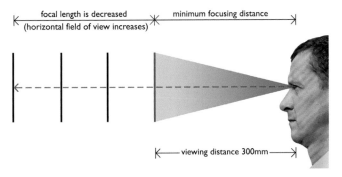

Fig. 74. By varying the distance at which the image is held, the focal length is changed using the eye instead of the camera lens.

If the image is now assessed on a desk from a standing position where our natural convention is to view the page as straight on as possible, this distance can be increased to as much as 600–700mm, and the single view cone of 39.6° is now wider than the image page itself. Because no clear instructions are given stating the importance of viewing the image with one eye from a fixed distance, the image is naturally viewed with both eyes at a greater distance, which considerably reduces the visual impact of the turbines. The degree of underestimation can be anything up to a factor of four on an A3 page depending on the field of view.

The problem here is quite simple. The distance between the eye and the image is used effectively to change the focal length of the image instead of using the focal length of the camera lens. The nearer the image is held, the narrower the field of view and the further away the image is held, the wider the field of view. This is a very unreliable and impractical method open to serious inaccuracies, as shown in the next two pages. This can be easily and more accurately overcome by using a combination of focal length and picture size applied to an image held at a natural and comfortable viewing distance. This is referred to in more detail in Chapters 9 and 10.

THE DIMINISHING TURBINES

The graphic on the next page illustrates the degree of possible underestimation if the image is incorrectly viewed from a greater distance. All the turbines are the same height and are rendered from the same viewpoint at different focal lengths. It can be seen that the wider the field of view, the greater the increase in perceived distance.

The first two turbines show the scale at 80mm and 75mm, which are two of the focal lengths which independent studies and the public generally find best represent what is experienced in the field from the actual viewpoint. The third turbine is shown at a focal length of 50mm, which is the standard or normal lens. By comparison, the last three images show a considerable reduction in scale if the turbines are shown within panoramic images with a field of view of 65.5°, 73.7° and 90° respectively, which are commonly used in windfarm visualisations on an A3 page.

If we compare their scale to the 75mm image, the reduction can be up to a factor of three if the field of view is 73°, and up to a factor of four if the field of view is 90°. This factor will increase if an image formed by cylindrical projection is viewed flat, because the scale of the turbines will decrease towards the edges of the panoramic image. This is further explained in Chapter 6, page 55.

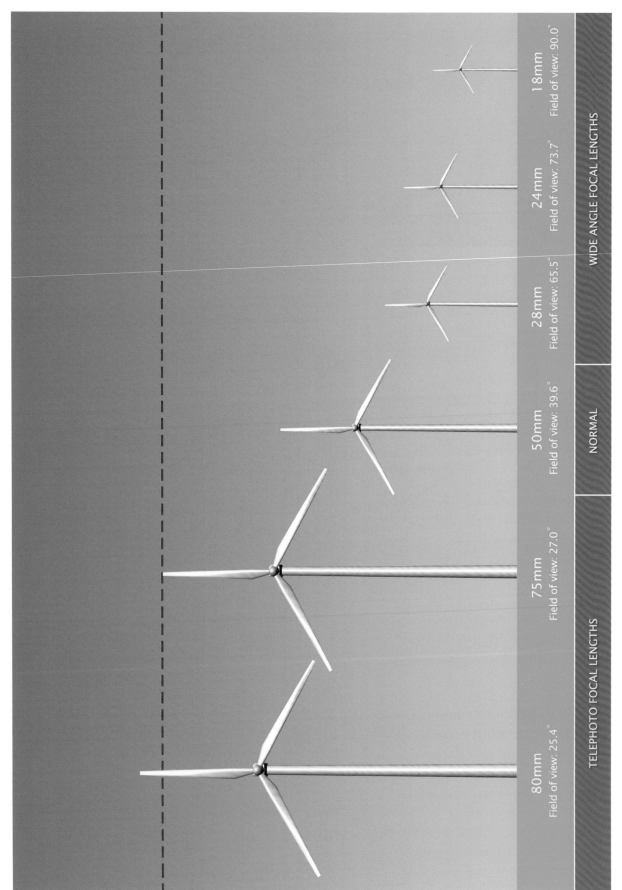

Fig. 75.

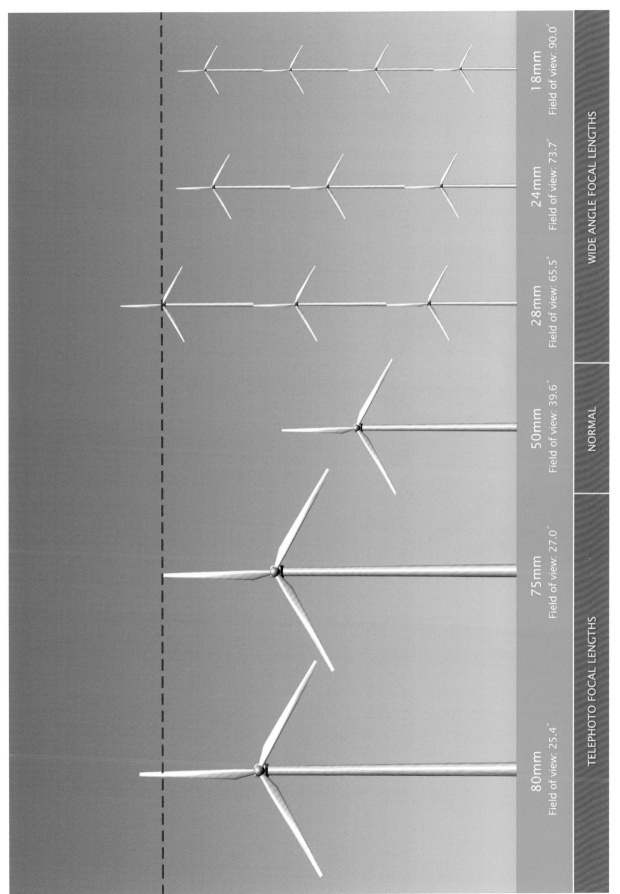

18mm
Field of view: 90.0°

24mm
Field of view: 73.7°

28mm
Field of view: 65.5°

50mm
Field of view: 39.6°

75mm
Field of view: 27.0°

80mm
Field of view: 25.4°

WIDE ANGLE FOCAL LENGTHS

NORMAL

TELEPHOTO FOCAL LENGTHS

Fig. 76.

COGNITIVE MAPS

Before concluding this chapter, it will help to briefly explain the importance of cognitive mapping for our everyday lives. Although we all have the ability to recall scenes or events, it is important to realise that we do not carry photographs in our heads, it is a much more complex system. Cognitive maps or 3D spatial images within our visual memory enable us to move around our environment. Without them, we would be completely lost. The original impetus for recent research on this subject was Kevin Lynch's book *The Image of the City* published in 1960. As a city planner, he hoped to improve the living environment by identifying paths, landmarks, nodes, edges and districts as the key components of recognisable places: in other words, to create identifiable spaces within which we can orientate ourselves.

Much research has been carried out since then and subsequent studies have shown that the more familiar a person is with their environment, the more accurate their cognitive mapping and actual mapping of an area becomes. Studies carried out by the Wellcome Department of Cognitive Neurology, University College London (Maguire et al., 2000 and 2006) found that the hippocampus where the brain keeps its maps or spatial representations of the environment was larger and more highly developed in London taxi drivers. This is due to the vast amount of detailed spatial navigational information they need to learn and retain, traditionally known as 'The Knowledge'.

Fig. 77.

Close your eyes for a few minutes after you read this paragraph and recall from your memory the route of a familiar walk or your journey home from work. You will quickly realise that we do not mentally envisage a series of still images, we are recalling spatial images or cognitive maps which we can mentally pass through. The more familiar we are with the environment, the more detail we can recall.

This is borne out by observations over many years of how local people responded to the visualisations in environmental statements when they were viewed in their local council facility. To them, the images appeared wrong and looked much further away. The images they were viewing simply did not match their own cognitive map of the familiar environment surrounding them, which was built up in their memory over many years.

As a photomontage is the *only* means by which the public can gauge visual impact, I have also found that when provided with a photographic montage of an appropriate focal length, they have a very clear understanding of potential impact of a windfarm development. Local people who are very familiar with their environment have as much ability to assess visual impact when presented with suitable images as the landscape professional who visits the area for a few days or weeks at the most and represents the interests of the developer. Local cognitive knowledge and perception can be invaluable in the assessment process.

In the case of windfarms, that assessment process is largely dependent on our understanding of photographic images. Over the next two chapters, I will explain how the images cannot be correctly viewed by a wider audience because of the technical and practical problems of the viewing methodology promoted by the Landscape Institute, SNH and others. Fortunately for the reader, only the next chapter involves some technical content, but hopefully presented in a comprehensible way.

6

THE TECHNICAL PROBLEMS

CREATING PANORAMAS

Paragraph B17 on page 175 of the SNH Guidance (2006) explains that while single frame photographs have a correct viewing distance using the principle of Leonardo's Window, it can also be applied to a panorama where the equivalent of Leonardo's Window would be a glass cylinder with the eye point in the centre. The correct viewing distance, it explains, is always the same radius as the cylinder.

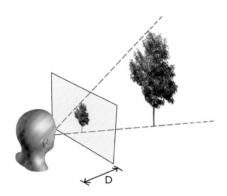

Fig. 78. A planar image can be superimposed on the scene it represents when viewed from the correct distance.

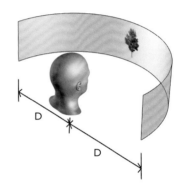

Fig. 79. A cylindrical panorama can be superimposed on the scene when viewed from the centre of curvature of a curved surface whose radius is the correct viewing distance.

Redrawn and adapted from figures B13 and B14 in Visual Representation of Windfarms: Good Practice Guidance (SNH 2006).

Fig. 80.

Because a panorama is made up of a series of overlapping photographs, the images have to be taken using a special panoramic head mounted on a tripod where the centre of the camera's front nodal lens or eye revolves around a fixed vertical axis. Put simply, all the photographs are taken from an exact fixed point, but each photograph will have a different perspective because the camera angle has been changed, so we end up with a multi-faceted image, as shown by the overlapping straight red lines in Fig. 81. The greater the number of facets, the greater the accuracy of the image.

By now curving the image using computer software, a process known as cylindrical projection, the overall image can be geometrically and optically formed into a curve, as shown by the broken black line in Fig. 82 which is the way the image should be viewed. Regardless of whether the image is a flat single frame or a curved panorama, the viewing distance is the same (see Appendix 4).

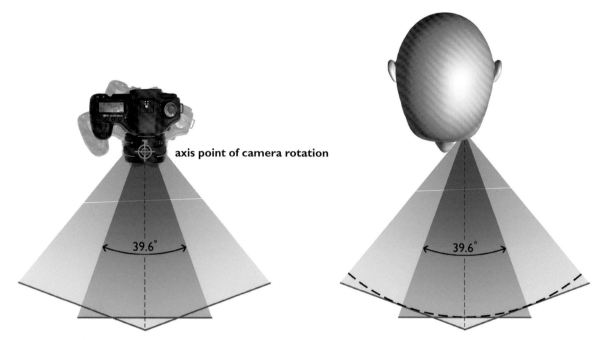

Fig. 81. Overlapping 50mm photographic images taken in 20° increments using a panoramic head.

Fig. 82. The flat photographic images can be transformed into a curved image using stitching software.

In contrast to the simple enlargement formula explained on page 24, the calculations for the viewing distances explained in Technical Appendix B of the SNH Guidance are much more complex. To the public, professionals and even landscape architects, they are difficult to understand and involve the application of mathematical formulae requiring the use of a scientific calculator:

For a single frame image:

$$d = \frac{w}{2 \tan\left(\frac{A}{2}\right)}$$

where d is the correct viewing distance in mm, w is the image width in mm, A is the horizontal field of view in degrees and tan is the trigonometric function.

For a panoramic image:

$$d = \frac{180\,w}{\pi\,A}$$

where d is the correct viewing distance in mm, w is the image width in mm, A is the horizontal field of view in degrees, and π has its usual geometric value.

The co-author of the SNH Guidance in his unpublished paper, *Photography and Presentation of Photographs for Environmental Assessment Work*, (Ian McAulay, TJP Envision Ltd, 2002) which is one of the papers referred to in the Guidance, states that the viewing distance is simply based on an enlargement factor of the original image width (36mm) relative to the size of the printed image. For no apparent reason other than to increase the technical complexity of the methodology, this simple calculation was replaced by the unnecessarily complicated formulae when the Guidance was published in 2006.

IMAGE PROJECTION

When photographs are overlapped horizontally to form a wider panoramic image, they are seamlessly joined together by stitching software. This can be created in two different ways: planar or rectilinear projection which creates an image which should be viewed flat, and cylindrical projection where the image should be viewed in a curve equal to the horizontal field of view of the panoramic image.

Both types of projection can be created almost at the touch of a button in stitching software, providing the images are made up of technically correct single frame images. Professional stitching software bases its calculations on the image metadata embedded within the original photographs themselves, so very accurate results can be achieved.

Planar (flat) projection

If the panoramas are to be presented as printed images to be viewed flat in an environmental statement, only planar projection should be used, because it maintains an accurate landscape profile. When the image is created, care must be taken to ensure that the stitched image does not exceed three overlapped 50mm images taken in 20° increments. Beyond three images, the landscape profile stretches towards the outer edges of the panorama because of the way it is calculated in stitching software to maintain the horizontal lines.

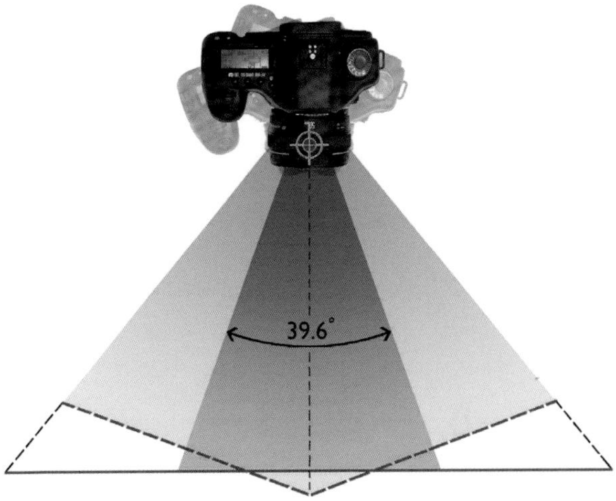

Fig. 83. Overlapping photographs with different perspectives taken on a panoramic head can be corrected by stitching software to form a planar image.

Cylindrical (curved) projection

This is the projection method most commonly used in panoramic windfarm visualisations, but because the curved image is viewed flat in an environmental statement, serious inaccuracies occur. Unless carried out under very specific studio conditions, cylindrical projection can only be correctly viewed in a single frame viewer on a monitor or projection screen where the distance from the point of projection is precisely controlled by computer to maintain the correct focal length throughout an arc of 360°.

Scanning the image horizontally through a viewer has a similar effect to turning one's head to scan the wider landscape as we would do in reality. Cylindrical projections are particularly useful in the assessment of large onshore, offshore and cumulative developments covering a very wide angle of view. This is referred to in more detail in Chapter 10.

Fig. 84.Cylindrical images can only be accurately viewed in a single frame viewer where the curved image has been geometrically corrected for viewing on a flat surface.

PERSPECTIVE DISTORTION AND COMPRESSION

Because distortion is not so obvious in a landscape analysis, it is best demonstrated in a location where there are many straight lines for visual reference, including an immediate foreground. If we look at the three images below, all the images have the same overall field of view of a 24mm wide-angle lens, which is 73.7°. The top two images appear very similar, but the bottom image is compressed horizontally and appears distorted.

Fig. 85. Image taken with a 24mm wide-angle lens cropped to a vertical field of view of a 50mm lens. The technically correct 50mm single frame photograph is outlined in red in the centre.

Fig. 86. Panorama formed by stitching three images together using planar projection. The fields of view are identical to image 1.

Fig. 87. Image formed by stitching three photographs together using cylindrical projection. Note that the field of view is now compressed. Because the stitched images have horizontally curved profiles, additional Photoshop infill work is required maintain the correct vertical field of view in the centre of the photograph.

The reason for this is because of the type of projection used. **Image 1** is a single frame, optically corrected, taken with a 24mm wide-angle lens and cropped to a vertical field of view of a 50mm lens. The 50mm single frame defined by the red outline fits exactly into the centre of the frame because both images share the same perspective.

Image 2 is a series of overlapping 50mm images using planar projection to create an image which should be viewed flat. Although the images all have a different perspective, professional software transforms the image into a single point of projection, similar to the first image. If this image is overlaid on **image 1**, the projection is very accurate and contains the same correct overall field of view of 73.7°.

Image 3 was created by cylindrical projection which is the method recommended in the SNH Guidance and the most commonly used in windfarm visualisations. The straight railings in the foreground appear to curve, and the scale of the buildings diminishes towards the outer edges of the image. This explains why straight roads, fences and gates curve downwards and power lines curve upwards in windfarm visualisations.

Although the image contains the same number of buildings, the actual field of view is reduced from 73.7° to 65.4°. Fig. 88 below explains why this occurs. Because the image should be viewed and only viewed from an exact distance while curving the image through an arc of 73.7°, the contained angle of view is compressed if the images are viewed flat. Even if the panorama is correctly viewed with one eye as shown in Fig. 89, the reduction of scale caused by cylindrical projection is further increased because the distance from the eye to the outer edge of the page is also increased by anything up to 20% on an A3 page.

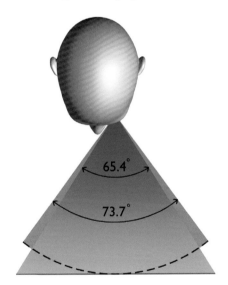

Fig. 88.

Fig. 89. If a cylindrical image is viewed flat, the viewing distance is increased towards the outer edges of the image.

Although the implications of this may not be immediately apparent, it enables windfarm consultants to fit a wider field of view within the same image width, further reducing the landscape scale. If we now compare **image 2** formed by planar projection with **image 4** formed by cylindrical projection, as shown in Figs. 90 and 91 on the next the page, the 73.7° field of view of the cylindrical image which is compressed into an angle of view of 65.4° is now increased to 86°.

Similarly, in a windfarm visualisation where the development extends over a significant part of a panorama or where the turbines are not central on the page, there will be a progressive reduction in the scale of the turbines from the centre of the image to the outer edges, reducing the visual impact of any turbines. The graphic in **image 5** shows the panoramic image formed by cylindrical projection superimposed over the same image formed by planar projection at the same image size. The only point where the two images match exactly is on the vertical centre line of the two photographs. It should be noted that in the remainder of the book, the angles of view may be rounded down for simplicity.

Fig. 90. The same panorama as image 2 on page 54 formed by planar projection.

Fig. 91. At the same image size, the image formed by cylindrical projection now contains a wider field of view because of the horizontal compression, and the vertical scale is diminished towards the outer edges.

Fig. 92. If the two images above are superimposed together, the degree of distortion due to cylindrical projection is clearly evident. The images only match on the centre line of the panorama.

In practice, windfarm visualisations are invariably viewed as flat images: they are viewed flat in environmental statements, on a computer screen or projected to a planning committee. Cylindrical projections should therefore not be used because they distort the landscape profile, diminish the scale of turbines towards the outer edges, and compress the overall field of view. Only planar projection, which can be accurately used up to, and just beyond a horizontal field of view of a 24mm lens, should be used, because it maintains an undistorted landscape profile. Horizontal compression is further referred to in Appendix 4.

THE VIEWING DISTANCE AND IMAGE HEIGHT

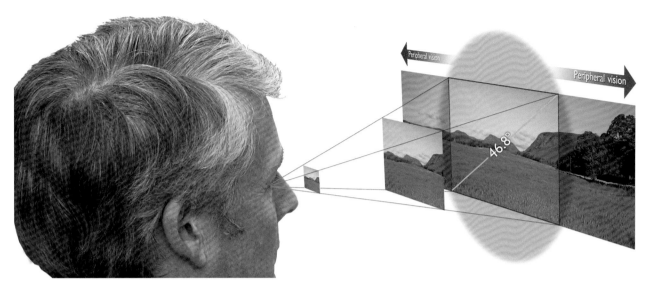

Fig. 93.

On page 24, in Chapter 3, we explained that the viewing distance is simply an enlargement of the focal length of the lens relative to the image size which can be calculated on its height or width. In a panoramic image, this is more difficult to calculate because the image is formed by extending the 50mm single frame horizontally to form a wider field of view which will be seen within the limited peripheral vision of one eye.

Fig. 94.

The panorama above using planar projection shows an overall field of view of a 24mm wide-angle lens. The width of the original 50mm single frame photograph cannot be readily identified because it forms part of a wider seamless image. As the focal length of any lens in the 35mm format is determined by its diagonal field of view within a 3 × 2 frame, the original single frame photograph will therefore occupy a 3 × 2 rectangle in the centre of the image. Because the extent of the horizontal field of view of the single frame is never defined, the only controlling dimension in calculating the viewing distance is its height. To now calculate the height required for a viewing distance of 300mm, which is the minimum requirement of the SNH Guidance, the following formula based on the enlargement factor of the original 24mm image height can be used:

300mm (viewing distance) ÷ 50mm (focal length) × 24mm (original image height)=printed image height
300 ÷ 50 × 24 = 144mm

However, a rather serious anomaly occurs here. Page 138 of the SNH Guidance (2006) states that the minimum image height should be 'over 130mm for hand held material'. This height was later increased to 140mm in the fourth draft of SNH's *Landscape and Visual Impact Scoping Issues for Wind Farm EIA* issued in 2008, yet both could still have the same viewing distance of 300mm and the same 50mm focal length. How is this possible? The answer to this can be found in Paragraph A28 on page 166 of the Guidance, which states: 'Examples of horizontal fields of view of a variety of focal lengths in conjunction with 35mm film (with a negative size of 36 × 24mm) are shown in Table 14. Both "round number" focal lengths and commonly available focal lengths are shown in Table 18. (Diagonal fields of view are included for completeness as some lens manufacturers quote this as the field of view of their lenses, but this figure is of little practical use).'

The latter statement raises further questions about the credibility of the Guidance because, as explained in Chapter 4, all fixed prime lenses are specified by their picture angle or diagonal field of view regardless of camera format. If the images are 35mm format, then an image height of 131mm would return a viewing distance of 272.9mm and an image height of 140mm would return a viewing distance of 291.6mm. Both distances are below the minimum viewing distance requirement of 300mm. The reason for this discrepancy lies in the SNH viewing distance formulae, which are based on the horizontal field of view of a 50mm lens (an angle of 39.6°), which is impossible to define in a seamless panorama. The two images below illustrate this problem.

Fig. 95.

Fig. 96.

The upper panorama is the same as the image on the previous page; the lower panorama has been vertically cropped. According to the calculations in the SNH Guidance, both images have a focal length of 50mm and the same viewing distance because the calculation when applied to the lower image does not take into account its height within the 3 × 2 frame ratio. Under this method of calculation, even a single line one pixel high would still have the same focal length and the same viewing distance, which is clearly technical nonsense. Although such a situation would not arise in an environmental statement, it does clearly illustrate the next problem: the practice of cropping images which is necessary in cylindrical panoramas and is particularly marked in image pages containing two photographic panoramas and a wireframe. Cropping a panorama made up of overlapping photographs in reality increases the focal length of the image, so the image should look closer.

If we compare the 50mm panorama in Fig. 95 with the panorama below in Fig. 96, which is actually a 75mm focal length, there is no apparent difference in our perception of distance because the horizontal fields of view are the same. If anything, the lower image appears to have a slightly more distancing effect due to its decreased height. How can this occur if cropping the image vertically increases the focal length, which should make the image look nearer?

This again is due to the SNH method of calculation. If we reconsider the same two images in Fig. 97 (albeit reduced) and define the 3 × 2 image at the centre of the photograph, it can be seen that the diameter of the lower image is smaller compared to the upper image, when in reality they would be the images projected by the camera lens onto the same size of electronic sensor. As explained on page 36, different focal lengths can only be accurately compared if the images are the same height. If the two images are now compared at the same height in Figs. 98 and 99 below, it can be seen that our perception of distance is different. The lower image now looks closer than the upper image because the horizontal field of view has been reduced.

Fig. 97. The upper image is the same 50mm panorama as shown on the previous page. The lower image has been cropped vertically to a 75mm panorama retaining the same horizontal field of view.

Fig. 98.

Fig. 99. Compared to the images in Fig. 97 at the top of the page, the above images show the difference in distance perception between the 50mm panorama and the 75mm panorama when the images are printed at the same height.

Due to this error, any viewing distance calculated on the image height and a focal length of 50mm will return incorrect results when the image has been cropped. The viewing distance or point of projection can only be accurately calculated if the recalibrated focal length is known, because as the vertical height of the image is decreased, the focal length is increased. The two images in Fig. 97 suggest that we would expect the viewing distances to reduce as the image gets smaller in height, but, this does not happen. If the images are cropped vertically, the viewing distance does not change.

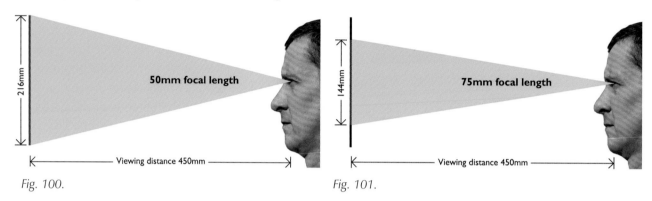

Fig. 100.   *Fig. 101.*

The image on the left above shows the viewing distance or point of projection for an image 216mm high with a focal length of 50mm. To calculate this, the following formula can be used:

216mm (printed image height) ÷ 24mm (original image height) × 50mm (focal length)= viewing distance
216 ÷ 24 × 50 = 450mm

If we now crop the same image to a height of 144mm, shown on the right, the focal length is increased from 50mm to 75mm. If we apply the same formula, the viewing distance remains the same.

144mm (printed image height) ÷ 24mm (original image height) × 75mm (focal length)= viewing distance
144 ÷ 24 × 75 = 450mm

It can therefore be seen that if a photograph has been cropped vertically, the viewing distance cannot be calculated on the image height, it can only be calculated on the comparative height of the original 50mm image because the focal length has been changed. As this can only be achieved by detailed computer analysis undertaken by experienced consultants, the visual representations cannot 'be based on a replicable, transparent and structured process, so that the accuracy of representation can be verified, and trust established' (Landscape Institute Advice Note, 2011). To establish whether or not an image has been cropped and the different implications on the calculation of the viewing distance just adds to the needless technical complexity, inaccuracy and impracticality of the whole viewing methodology. This practice not merely invalidates the stated focal length, it subverts the purpose of a 50mm reference standard. Examples of correct monocular viewing distances relative to image heights are shown in the table below.

Lens focal length	Viewing distance	Image height
50mm	300mm	144mm
50mm	350mm	168mm
50mm	400mm	192mm
50mm	450mm	216mm
50mm	500mm	240mm

Fig. 102.

Fig. 103.

PORTRAIT ORIENTATION

A further error occurs on page 192 of the SNH Guidance. Figure E4, similar to the image on the left, shows that panoramic photographs can also be taken in portrait orientation by turning the camera sideways on a panoramic head. From a technical point of view, this is not a problem if the image is a single frame image because the diagonal field of view which defines the focal length remains the same. However, serious problems arise if images in portrait orientation are used to form a much wider panorama.

By rotating the camera from landscape to portrait orientation, the vertical field of view is increased from 27° to 39.6°. If the 3 × 2 frame ratio is now defined within the centre of the image, the focal length is changed to 33mm, which is slightly wider than a 35mm wide-angle lens, as illustrated in Fig.104. The correct 35mm and 50mm fields of view are inserted to illustrate the problem. According to the SNH Guidance, this would still qualify as a focal length of 50mm.

Fig. 102 shows a 35mm format camera mounted in portrait orientation on a panoramic head. Fig. 103 shows the 3 × 2 image and the diagonal field of view. Fig. 104 shows a panoramic image with a horizontal field of view of a 28mm lens and a vertical field of view greater than a 35mm lens. The correct 50mm images are outlined in red.

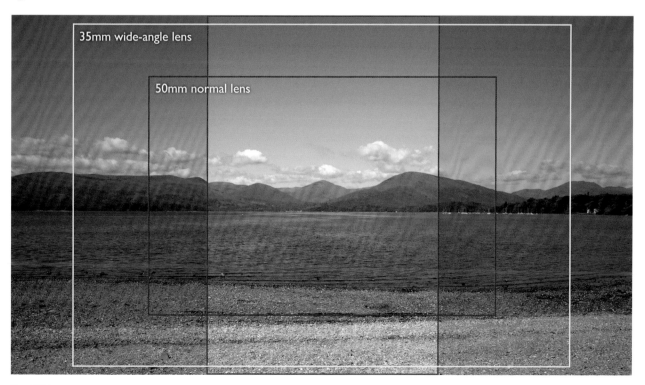

Fig. 104.

SNH GUIDANCE VIEWING DISTANCES

Because the science shows the SNH Guidance (2006) to be technically flawed, the viewing distances based on a focal length of 50mm in all the examples throughout the document are also incorrect. The most serious error occurs in paragraph 156, page 8, which states that 'the following photographs (Figures 22a–22c) of the existing Dun Law windfarm show a comparison of effect using alternative lens length for an image of the same size but varying viewing distance'.

Figure 22b: Photograph of existing windfarm taken with 50mm lens

Photograph taken using a 50mm focal length lens on a digital SLR. Horizontal field of view 25.9 degrees. Correct viewing distance 82.6cm. 26km to windfarm.

Photograph taken using a 28mm focal length lens on a digital SLR. Horizontal field of view 44.7 degrees. Correct viewing distance 46.3cm. 26km to windfarm.

Photograph taken using a 135mm focal length lens on a digital SLR. Horizontal field of view 9.7 degrees. Correct viewing distance 223cm. 26km to windfarm.

Reproduced from Visual Representation of Windfarms: Good Practice Guidance *(SNH, 2006).*

Fig. 105. For copyright reasons, the actual photograph has been replaced by a similar image. However, it is not the photograph that is of interest: it is the technical information on the page which is reproduced above and enlarged for clarity.

The photographic pages in question are only available in the printed version of the Guidance and not downloadable from the SNH website. They are shown overlapped above so the technical information on the bottom two pages is visible. It should be emphasised that the pages are A3 format with an image size of 380mm × 244mm and the focal lengths shown are 28mm, 50mm and 135mm. If we concentrate first on Figure 22b, the top image, it states that the photograph was 'taken using a 50mm focal length lens on a digital SLR. Horizontal field of view 25.9 degrees. Correct viewing distance 82.6cm.' The horizontal field of view for a 50mm lens is 39.6°, not 25.9° as stated. As a result, the viewing distance is seriously inaccurate.

As mentioned on page 24, the viewing distance for a single frame image can be calculated by simply using an enlargement factor of the image height or width. The image does not have the correct 3 × 2 ratio of the 35mm format, so the calculation will have to be based on the image width. If we now use the following calculation:

380mm (printed image width) ÷ 36mm (original image width) × 50mm (focal length) = viewing distance

380 ÷ 36 × 50 = 52.77cm (not 82.6cm as stated)

The horizontal fields of view for the 28mm focal length and the 135mm focal length are stated as 44.7° and 9.7° when they should be 65.5° and 15.2° respectively. Applying the same enlargement formula above, the viewing distance for the 28mm image is 29.5cm, not 46.3cm as stated and the viewing distance for the 135mm lens is 142.5cm, not 223cm as stated. This in turn gives a viewing distance error of 56% for all three focal lengths. The reason for this discrepancy is because the SNH viewing distances have been calculated using their single frame formula on page 52 which is based on horizontal fields of view which are incorrectly stated.

The incorrect viewing distance calculations are not limited to the single frame images. For anyone with the printed version of the SNH Guidance (2006), they also apply to all the A2 size panoramic images referred to in Table 15 on page 130 which are folded into clear plastic pockets within the Guidance document (these images are not available in the downloadable version of the Guidance). Because the width of the original single frame image is not defined within the overall panorama, the only controlling dimension for calculating the viewing distance is the image height. Applying the enlargement formula on page 24, it will be found that none of the viewing distances are correct.

Fig. 106.

This can be confirmed by a simple test. Measure the height of the printed image on the page, divide it by the height of the original image (24mm) and multiply by the focal length (50mm). The viewing distance will be different to that stated on the page, so the accuracy of the images cannot be verified. The reason for this is because the focal lengths are not 50mm, but there are other inconsistencies in the technical data which make it impossible for the images to be verified and open to scrutiny. Even computer analysis of the images returns different viewing distances depending on which data are given priority.

Fig. 107.

It should also be noted that nowhere on the image pages does it state that visualisations must be curved and viewed with one eye for correct perspective viewing. As the majority of people are likely to read the SNH Guidance (2006) in its electronic version where the larger images are not downloadable, try measuring the height of any panoramic image in an environmental statement. The chances are the viewing distance will be incorrect if the stated focal length is 50mm. In the hundreds of visualisations I have investigated since the publication of the SNH Guidance (2006), almost all had incorrect viewing distance/focal length information, invalidating the accuracy of the viewing methodology.

7

THE PRACTICAL PROBLEMS

TESTING IMAGES IN THE FIELD

The generic difficulties with viewing images in the field were revealed through tests carried out in conjunction with The Highland Council. Because no explanation was given on any of the visualisation pages that the images should only be viewed with one eye and curved in an arc equivalent to the field of view, when converted to transparencies, they did not align horizontally with the landscape when viewed as a flat image at the specified distance. When viewed with both eyes as we would normally view a photograph, the landscape was seriously out of alignment and scale.

Fig. 108. Empirical testing in the field applying the principle of Leonardo's Window revealed serious problems with many of the visualisations because of distortion, the degree of curvature required and the inability to properly focus on the image. If viewed as a flat image, the transparency did not remotely fit the landscape profile.

Even when held at the specified distance and viewed with one eye, some images had to be curved in a parabolic arc of up to 180° to fit the distance cues of the landscape, which made focusing across the image impossible because the outer edges were too near the eye. In almost all cases, the immediate foregrounds did not match up with middle distance and long distance cues, causing serious parallax problems. Without a semi-rigid cardboard frame containing the transparency which could be bent to the required arc of view, it was also found to be impossible to curve the image correctly even in light winds and the slightest variation in curvature seriously distorted the landscape. These problems are further compounded with printed images because the distance cues which enable us to align the transparency with the real landscape are no longer visible and are effectively blocked out by the wide panoramic image.

Even if clear viewing instructions were given, whilst we all have the ability to roughly gauge a defined distance perpendicular to our line of sight, it is quite another matter to gauge an exact distance and curvature away from the eye. It is also surprising to find that many people cannot effectively close one eye.

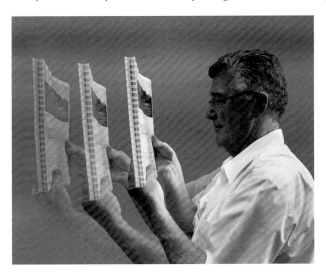

Fig. 109.

Without the aid of some sort of specialist apparatus, I have found over many years that when a heavy environmental statement is hand-held in the field, it is impossible for anyone to hold the page at an exact distance and curve the image in a precise arc equivalent to the field of view while viewing the image exactly perpendicular to the point of projection using one eye only. Because the focal length of the human eye is only around 22.3mm, the distance from the exact centre of the image to the eye requires sub-millimetre accuracy. This level of accuracy is simply not possible or achievable, which further discredits the whole practicality of the methodology. For a member of the public or even a non-landscape professional, none of this is even explained, so the images can never be viewed correctly.

The Landscape Institute Advice Note states that 'panoramic images should be curved so that peripheral parts of the image are viewed at the same intended viewing distance, or viewed by panning across a flat image with the eye remaining at the recommended viewing distance'. The importance of this viewing distance is also emphasised in the *Letter issued to all Heads of Planning* by the Chief Planner, Scottish Government (Mackinnon, 2009). Because there is only one exact point at which an image can be correctly viewed when applying the principle of Leonardo's Window, the image cannot be viewed by rotating our head as we would do in reality to scan a wider field of view in a landscape. This is because the rotational axis of our neck is not on the same rotational axis as the lens of our eye, so maintaining a precise viewing distance across the image is impossible.

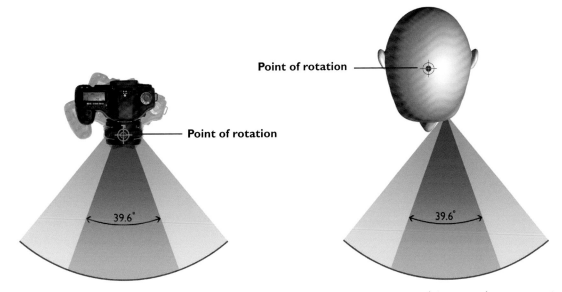

Point of rotation

Point of rotation

39.6°

39.6°

Fig. 110. When a panoramic head is used to create a series of overlapping images, the eye or front nodal lens of the camera rotates around same vertical axis.

Fig. 111. Because our neck is not on the same rotational axis point as the camera lens, an even viewing distance cannot be maintained across the image.

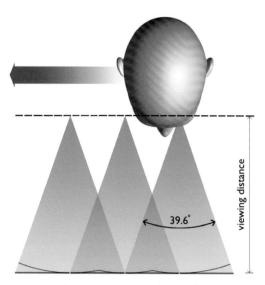

Fig. 112. The image projected onto the retina of the eye is now made up from several different curved images and points of projection which just appear as an overall distorted image.

If the photomontage is viewed on a flat plane, which is the other viewing option, there are additional problems. Because a cylindrical image has a single point of projection, which is the centre of the imaginary glass cylinder, when viewed flat the image is made up of many different points of projection, each with its own cylindrical image. It now has to be viewed by panning the eye across the panorama while maintaining the correct viewing distance. Apart from the fact that it cannot fit the landscape because it is horizontally compressed and distorted, the retinal distortion is such that accurate viewing by this method is not possible.

The problem is increased considerably if the image is viewed with two eyes because our peripheral vision is increased. Try closing one eye, and then open both eyes. You will see that our peripheral field is much wider when viewed through two eyes. Because the curved image is now viewed flat, the viewing distance is also incorrect. As the viewing formula in the SNH Guidance is based on the horizontal field of view of the single frame photograph in the centre of the panorama, which is now compressed from 39.6° to 38.2°, any '50mm image' extracted from a panorama will have cylindrical distortion, a different focal length and contain no metadata for verification.

PUBLIC ACCESSIBILITY

Access to suitable visualisations has always been a problem for the public. The non-technical summaries which are available free of charge are generally nothing more than advertorial leaflets promoting the benefits of green energy, often containing no visualisations, and where these have been supplied, they are devoid of any technical information and so misleading that any form of meaningful on-site assessment is impossible.

Until recently, because the main environmental statements cost several hundred pounds, the only way the public could view the visualisations was in their local council facility where the documents had to remain for the duration of the consultation period. As a result, it was generally not possible for anyone to assess the images on site. It would be logical to assume that as the bulk of the information in the heavy environmental statements is written material for professional assessment, the visualisations could have been incorporated into a separate A3 document and made available to the public at a reasonable cost, but this was generally never done.

Although environmental statements are now available on CD-ROM at a much cheaper price and therefore more accessible, the situation has become even more confusing and misleading. Many CDs are unnecessarily difficult to navigate and the photomontages which are supplied electronically require the images to be printed at A3 size or above. Extended A3 images are now also becoming very common. Access to visualisations is now also possible through most local authority e-planning portals; however, the resolution is quite often of poor quality and downloading times can be very slow even with broadband.

Fig. 113.

As most people only have an A4 printer for their everyday needs, printing is not possible without going to a professional output centre, which is costly, and so easy access to on-site assessment is compromised. The only way they can readily assess the images is on a computer monitor, where a viewing distance cannot be accurately applied, nor can the screen be curved for correct perspective viewing.

Fig. 114. A viewing distance cannot be applied to an image on a computer monitor, nor can the screen be curved.

Fig. 115. Similarly, a viewing distance cannot be applied to an image projected to a planning committee.

This important point also applies to a panoramic image projected to a planning committee, who are also members of the public. As a viewing distance cannot be applied to a projected image, committee members have no option but to view the much wider image, which under-represents the visual impact and distorts the landscape. Even if large panoramic prints were made easily available to the public, they would be of little practical use. Studies and tests carried out in conjunction with The Highland Council have found that visualisations in environmental statements exceeding A3 formed by cylindrical projection which have to be held in an exact curve are impractical for use on site in a light breeze. Even under totally calm conditions, it was not possible to view the visualisations in a precise curve because of the fold in the image page.

Fig. 116. Even in a light breeze, the images printed on paper could not be curved in an arc equivalent to the overall field of view, making the application of an exact viewing distance impossible to achieve.

In August 2008, I led a deputation of four Scottish councils, two Scottish national parks and a representative from the English national parks to meet the Scottish Government to present a 'Case for Change' regarding the presentation of visualisations in environmental statements. The meeting was also attended by other government bodies, SNH and the co-authors of the present Guidance.

When the practical problems of viewing such large panoramic images in the field were pointed out, the co-author of the SNH Guidance demonstrated how professionals do it. As illustrated below, using a combination of spread fingers on both hands to attempt to gauge the correct viewing distance, the image was viewed flat with both eyes and scanned sideways maintaining the hand combination over its entire length and viewed perpendicular to the page. Apart from the image being obscured by the combination of hands extending from the bridge of the nose to the image, the principle of Leonardo's Window can no longer apply if the image is viewed with both eyes. Even if the image was viewed with one eye, the limited peripheral vision would be further restricted on one side by the hand combination.

More importantly, regardless of whether the 50mm panorama was viewed with one eye or two eyes, the retinal distortion due to cylindrical projection would be unacceptable on the flat page and the scale of the turbines would be under-represented because of linear perspective. In reality, such a viewing method would involve taking some form of table device to site. Purely from a practical point of view, the sheet would have to be pinned down to prevent any wind uplift and stretched flat to prevent any further distortion caused by the folds in the image sheet.

Fig. 117. How professionals do it. Using a combination of spread fingers to establish the viewing distance, the image is then viewed by panning across the flat image.

It would seem to be apparent that it is an extremely cumbersome and impractical solution compared to simply viewing a single frame image in the real landscape. The letter issued by the Chief Planning Officer for Scotland to all heads of planning in January 2009 (Mackinnon, 2009) which resulted from the meeting in August 2008, although constructive in other aspects, endorses this present viewing methodology without perhaps fully understanding the ramifications of what was presented at the time.

8

THE FUNDAMENTAL ISSUE

From the foregoing it is clear that the viewing methodology used in current practice is seriously flawed. It is not good science, not good mathematics, totally impractical and, it might seem, has been created to deliberately mislead. The fundamental issue here is not about whether the correct viewing distance is 285mm or 300mm, whether the image height is 130mm or 140mm or even whether the image should be curved or flat and viewed with one eye or two eyes. The problem is that the one single visual format is being used to encompass all forms of assessment. An unnatural distance between the image and the eye is being used to effectively change the focal length instead of using the camera lens itself. While it is accepted that it may be necessary to show a windfarm in a wider context for landscape assessment, the same image should not be used for the related but separate assessment of visual impact, when it requires the application of a complex viewing technique which is open to misinterpretation, inconsistencies and serious inaccuracies.

The human visual system is much more complicated than a photographic camera. Although the eye has certain similarities, what we see has evolved over many years of learning and experience. We do not see the two-dimensional image projected onto the retina of the eye because it has already been processed by the brain. When we view the three-dimensional world around us, we are constantly gauging distance which we interpret by using the shallow depth of field and blur along with the information sent to the brain from the eye muscles. In other words, we take distance into account when we re-calibrate the scale of more distant objects by size-constancy scaling, as explained in Chapter 4.

A photograph taken with a 50mm lens will under-represent the perception of distance and vertical scale of more distant objects because we are looking at a flat image containing no depth information. The principle of Leonardo's Window can only be accurately applied to a transparency where we have the distance cues to scale the retinal image to the real world: it cannot be simply reversed to apply to an opaque photograph where everything lies on the same focal plane and the distance cues are no longer visible. While there is a theoretical viewing distance for photographs, it is largely irrelevant because we naturally view images with both eyes at a comfortable distance relative to the size of the page. The problem with panoramic images formed from multiple 50mm frames is that they are designed to be viewed at a much closer distance with one eye, which inevitably leads to an under-representation when viewed on an A3 page at a more comfortable and natural distance.

Current guidance and advice is over-complicated, contradictory and devoid of prescriptive foundations, while the related research and opinion on the subject has been and continues to be largely ignored. The viewing methodology is supported by selectively cherry-picking from unrelated applications while ignoring other directly related facts and issues exemplified in Technical Appendix C of the SNH Guidance (2006), which references Professor Gregory's *Eye and Brain* regarding the acuity of the human eye, yet ignores the important difference between linear perspective and size-constancy scaling which is explained in detail in the same publication. The methodology also contains a number of serious technical problems, resulting in images which are highly distorted and compressed, focal lengths which are rarely accurate, and unusable viewing requirements which are invariably incorrect.

The problems faced by the user are also considerable: not only is the recommended viewing method impractical and unreliable, the images themselves are not readily accessible in the correct format. Whatever the semantics of professional landscape assessment vocabulary, two points are clear: prediction and evaluation are at the heart of any EIA, and visual tools provided as part of a landscape and visual impact assessment must serve a number of purposes and the needs of many different audiences. It is evident that the visualisations as presented in environmental statements do not provide safe prediction of visual impact, nor is the applied viewing methodology technically acceptable for use by a non-professional and public audience.

It was always my understanding that the panoramic photomontages were for landscape professionals, but this has proved not to be the case. It is now evident that landscape architects acting on behalf of windfarm applicants do not use photomontages within their process of assessment: they base their assessment on the on-site experience using the wireframe images as a reference aid. This has also become clear in recent evidence given by landscape professionals at public inquiries, who on the one hand emphasise the importance of the viewing distance, yet clarify that they do not use the methodology themselves. This is further endorsed by a report (Scottish Natural Heritage, *Visual Representation of Windfarms Good Practice Guidance Review 2011–12, Scoping workshop* [Turnbull, 2011]) prepared for the Landscape Institute Technical Committee, which states, 'Landscape professionals may use wireframes and other visual aids during their assessment but in the (sic) no circumstances do the actual photomontages form the basis for assessment …'

The then Chairperson of The Landscape Institute Scotland Branch also wrote in her consultation response to the Draft SNH Guidance (*Visual Analysis of Windfarms: Good Practice Guidance – Consultation Comments*) in 2005: 'I do not believe anyone really uses the viewing distance in practice (no one puts these in a frame and measures the distance from the eye to the page and so on), particularly as you have the real landscape to work with.' This view is also shared by many others. A situation therefore exists where the general public, town planners and decision makers who have no understanding of the specialist viewing techniques are expected to be able to make an informed decision from photomontages which seriously distort landscapes and under-represent visual impact, while the same visualisations are not even used by landscape professionals themselves. The question therefore has to be asked: who are these visualisations for?

For over sixteen years, the application of a viewing distance methodology has been the main cause of the widespread complaints of visual misrepresentation. For the wider audience and even many independent landscape professionals themselves, the current visualisation methodology is not understood, inaccurate, highly misleading and cannot be properly applied in practice. These same problems have also been recorded by inquiry inspectors. In a report for a 12-turbine development at Hall Farm, Routh near Beverley, Yorkshire in 2009, the inspector stated, 'Photographs have been provided to illustrate the landscape setting of the Minster and photomontages to show the degree of visibility of the proposed turbines. I treat these with caution. This is because my perception of the Minster from the Westwood is wholly different from that conveyed in the images presented', and he went on to state, 'My visits to existing windfarms elsewhere have been informative and confirm my opinion that the reality of the proposed situation is not accurately captured from studying the photomontage images.'

It is for these reasons that clear standards and a proper understanding of the main requirements of photomontages are so necessary. The starting point must be a full understanding of their purpose, the audiences who will use them and the way in which the images will be viewed. Professor Stephen Sheppard in his chapter 'Validity, Reliability and Ethics in Visualizations' in the 2005 edition of *Visualization in Landscape and Environmental Planning*, identifies three fundamental objectives: to convey a clear understanding of a proposed project, to evoke unbiased responses to the proposed project and to demonstrate credibility of the visualisations themselves. He goes on to say that in the absence of recognised or formalised standards, he proposes the following six principles: 'Accuracy, Representativeness, Visual Clarity, Interest, Legitimacy and

Access to Visual Information'. He also expresses his concern as follows: 'With the steadily increasing access to more user-friendly software, the lack of training or guidance in the use of visualization poses a significant threat to valid public processes. The heavy-reliance on imagery to sell market-driven products makes it inevitable that deliberate distortion for commercial purposes will be attempted in planning, too. One of the most urgent needs then is to develop widely recognised ethical guidelines for landscape vizualisations'.

I totally endorse Professor Sheppard's call for a code of ethics and the need for recognised or formalised standards. It is vital that all windfarm visualisations which are submitted to any planning process are capable of verification by local authorities before they are subject to any public consultation. In order to achieve this, there must be fixed and verifiable standards. In the next section of the book, I put forward a few basic recommendations which have been proved and tested over time which will lead to a much clearer understanding of potential visual impact for all audiences.

Because of the lack of any prescription, we are now in a situation of having different presentation formats which, according to present guidance, all have the same focal length of 50mm, yet they appear very different. Photographically, only Fig. 118 is correct. The 50mm image is outlined in red in Figs. 119 and 120.

Fig. 118. The 50mm image showing the 3 × 2 ratio of the 35mm format.

Fig. 119. 50mm focal length with a horizontal field of view of a 24mm wide-angle lens (73°). The photographs are shown in landscape orientation.

Fig. 120. 50mm focal length with the same horizontal field of view as Fig. 119. The photographs are shown in portrait orientation.

Several people have suggested that it would be useful to include case studies comparing applicants' planning visualisations with the constructed reality. Such studies cannot currently achieve a great deal because regardless of whether images are photomontages or photographs of the real scene, the same perceptual problems will occur over which presentation format best represents what we see. Assessments could only be made by visiting the viewpoints and comparative illustrations are further obstructed by copyright issues.

If the original photomontages were accurate there should be no discernible difference between the images when shown 'like for like' in their respective formats. The main problem lies in the different way the images affect our perception of distance. It is therefore not possible to judge the 'representativeness' of visualisations until fixed standards are established to provide a solid and reliable basis for comparison.

THE MOON ILLUSION

Why does the moon look smaller when it is high in the sky than it does on the horizon? It has mystified people for centuries and even today, several different explanations are still put forward on the internet. However, if we take photographs using the same lens, the moon is actually the same size regardless of its location. If we also view the moon just above the horizon with one eye through a small peephole, the scale of the moon appears to be smaller than that seen by the other eye. Why should this visual phenomenon occur?

Fig. 121. The centuries-old question. Why does the moon appear larger on the horizon than it does when it is high in the sky?

According to Professor Knill of the Center for Visual Science, there is only one logical explanation: it is due to incorrect size scaling caused by a misperception of distance (personal comment). When we look at the moon high in the sky there is no visual context. The only way we can judge distance is by using stereopsis – the difference in distance between our two eyes. When we focus on an object, our eyes converge on distances up to ten metres. Beyond that, our lines of sight become parallel and everything appears to be more or less at the same depth. Because our visual system is not capable of recognising cosmic distances we are effectively looking at a flat two-dimensional image. Our eyes tell our brain that we are looking at an object which must be over ten metres away and because of its small size on the retina of the eye, the moon must be small, regardless of what our intellect may tell us.

However, when we view the moon on the horizon, we have many familiar perspective cues. We know by experience that the horizon is a few miles away and can see that the moon now lies beyond the horizon line as it rises. When our brain does its scaling, the moon has to be a lot bigger because it appears to be further away.

This demonstrates that regardless of the size of the retinal image, our ability to correctly scale the size of an object is dependent on our perception of distance in the real world.

SECTION 2

9

THE WAY AHEAD

INTRODUCTION

Before defining recommendations for a way ahead, it is important to focus again on the purposes of photo-montage visualisations, their limitations and the audiences who will use them. In the case of windfarms, the purpose of a visualisation is threefold: to illustrate landscape change, to provide an accurate representation of the predicted scale and appearance of the proposal from the same viewpoint, and to apprise a wide audience so that informed judgements can be made. They can also be used in the early stages of any project as an effective tool to aid development design.

Photomontages will always have limitations because a two-dimensional photograph can never totally represent what we see in reality, particularly for a landscape where considerable distances are involved. However, photography is the only practical means available and if properly defined, it can be used to provide realistic predictions. The principle of accuracy is sometimes disputed and can be reflected in the photomontage's truthfulness or honesty. In windfarm visualisations, it is justified purely in terms of correct perspective geometry regardless of whether or not it bears any resemblance to the viewer's perception of what is actually seen from the viewpoint. The principle of accuracy should apply not only to the technical aspects of photomontages, it should equally apply to how well an image represents our visual experience so that the impacts of all developments are known and understood prior to any planning decisions.

Professor Stephen Sheppard in his 1989 *Visual Simulation: A User's Guide for Architects, Engineers, and Planners*, states that 'a fully accurate simulation shows a view of the project that is not significantly different in appearance from the real view when seen from the same viewpoint'. In the case of windfarms, however, inaccuracy or dishonesty will sometimes not be fully apparent for some years. The construction time delays and the provenance of images created or funded by parties who derive considerable benefit from planning approval, make it doubly important that visualisations can be properly and easily verified at the planning stage.

The audience who will use photomontages is extensive. The smallest sector is landscape professionals, who are currently responsible for their production but who frequently assert that they do not use them in their professional landscape and visual impact assessments (LVIA). The greatest audience is encountered in the planning system for which any LVIA and associated illustrative material are ultimately produced. This sector is wide and diverse: a variety of professionals, consultees, planners, decision makers and the wider general public, who are all involved in understanding, commenting on or deciding upon the proposal depicted. The main priority is therefore the audience within the planning system rather than the landscape professional who prepares or assesses an LVIA submitted by applicants.

While writing this book I have often reflected on the huge challenge which faced John Harrison, the brilliant 18th-century clockmaker who ultimately solved the problem of calculating a longitude position for vessels at sea. A solution which was practicable was the overriding condition he had to meet in order to succeed. What is practicable is also central to the search for a visualisation solution, which must include the elements of familiarity and accessibility for a universal audience of all skill levels. Windfarm visualisations have to specifically fulfil a number of purposes covering a range of landscape disciplines, the needs of the planning system and at the same time, be adaptable to a range of individual project requirements. They also should be verifiable, accurate, not open to misinterpretation, clearly understood by a universal audience and accessible to all for comparison in the field. It is a very exacting specification with diverse and often conflicting requirements which cannot possibly be contained within a single visual solution.

To provide a solid foundation which is capable of verification by others, some prescription is necessary and should be applied to the fixed standards proposed in the following pages. Beyond these basic requirements, there is flexibility so that landscape professionals can make judgements on the optimal field of view, which may vary from project to project, and similarly, planners can request additional visualisations and techniques appropriate for a particular project under consideration. It is to be hoped that greater flexibility may encourage more practical innovation in visual simulation solutions. The starting point must be what is comfortable and practical for the viewer.

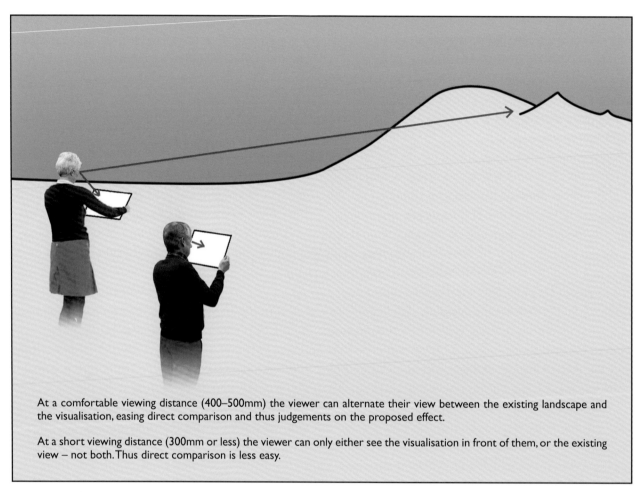

At a comfortable viewing distance (400–500mm) the viewer can alternate their view between the existing landscape and the visualisation, easing direct comparison and thus judgements on the proposed effect.

At a short viewing distance (300mm or less) the viewer can only either see the visualisation in front of them, or the existing view – not both. Thus direct comparison is less easy.

Fig. 122. Redrawn and adapted from Figure 35 in Visual Representation of Windfarms: Good Practice Guidance *(SNH 2006).*

IMAGES FOR COMFORTABLE VIEWING

Figure 122 shows two people viewing the same scene but using different viewing methods to assess the landscape. The woman is holding the image at a comfortable arm's length, while the man is viewing the image at a much closer distance. Why is this necessary?

Fig. 123. The woman is holding the image at a comfortable distance and comparing it to the real landscape.

Fig. 124. The man has to view the image with one eye at a much closer distance.

The woman is holding the image at a natural distance, which enables her to compare the photograph to the real landscape, 'easing direct comparison and thus judgements on the proposed effect'. The man, however, has to view the image with one eye from an exact fixed distance so that the centre of the image is perpendicular to his line of sight. He can only see the visualisation in front of him, or the existing view – not both, 'thus direct comparison is less easy'.

When we view a photograph, what is a comfortable distance? If the image is to be hand-held, it is somewhat limited. It is difficult to focus on an image closer than 300mm and this is particularly the case as our eyesight deteriorates with age. We are further limited by the length of our arms, so size becomes important because the larger the image, the further away we naturally hold it. Professor Benson identified that small photomontages were the main cause of underestimation and observed that for visual impact assessment, a full-page single frame image with a minimum height of 200mm gave the best representation of reality. He also recommended that what is comfortable for the viewer should dictate the technical detail and not vice versa.

The nearer an image is held, the more our eyes have to converge or angle inwards to focus on the same object. When we view an image in an environmental statement we are not simply viewing a photograph: we are using the photograph to assess scale by comparing it to the real landscape, so the further away we hold the image, the easier it is to accommodate the wide range of focusing distances. An image held at a comfortable arm's length therefore makes this assessment easier. Because of its size, an A3 print is naturally viewed at around the same distance, but the problem here is that while a 50mm image may be representative of what we see in terms of our in-focus vision, it under-represents landscape scale because of linear perspective.

How can this be overcome? From a very early age, we know that a picture does not represent what we see and that we have the ability to view it from a variety of different angles and distances. With the widespread popularity of small and convenient compact digital cameras and even camera phones, whether we understand the principles of photography or not, we know that zooming in or out on a subject gives us the impression of being nearer or further away without moving from the spot. It is logical therefore that at some point within the zoom, there is a focal length which reasonably represents what we see in a landscape situation within our experience and understanding of photographic images.

If the vertical scale in a photograph is similar to what we see at our point of focus in a landscape, we have a reliable impression of scale. It is similar to a technical drawing, but instead of viewing a two-dimensional scale which we associate with a flat image, we effectively use a three-dimensional scale when we view a landscape. By increasing the focal length of the camera lens, a realistic representation can be achieved. In most landscape situations, a focal length in the range of 75mm provides very satisfactory results. When viewed as a single full-frame image at a natural and comfortable distance on an A3 page, photographically it is the nearest to the scale we see in reality within the obvious limitations of photography. This is backed up by my own findings, independent reports, extensive research by The Highland Council, and the focal length perception study, *The Effect of Focal Length on Perceptions of Scale and Depth in Landscape Photographs* (2012) undertaken by the University of Stirling.

SINGLE FRAME IMAGES

Single frame images are the most economical to produce. This was the advice given to the SNH Steering Group by their spatial data analyst during the formulation of their Guidance. However, despite this advice, SNH and the windfarm industry have always resisted their use, although the SNH Guidance (2006) does make an allowance in their flow diagram, Fig. 37, which states that single frame images are acceptable providing that the field of view is able to show the key characteristics of the visual resource. If the purpose of an image is to give a realistic impression of the turbine scale, why is it necessary to show the key characteristics of the visual resource when this is already shown in the panoramic images which aid landscape professionals in their assessments in conjunction with their on-site experience? Additionally, all guidance and environmental statements emphasise that visualisations are tools to be used in the field to aid assessment and the Landscape Institute Advice Note states that images for impact assessment 'must include inspection at the location where the photograph was taken'. Why, then, is it essential to show the visual resource when a photomontage containing the extent of the windfarm can be viewed within the wider context of the real landscape? This is the very point made by the former chairperson of the Landscape Institute Scotland Branch.

The current lack of any prescription was clearly illustrated in an investigation into the visualisations of four windfarm applications near the southern boundary of the Exmoor National Park in Devon within relatively short distances of each other (Architech, 2008). Each application had different viewing distances, fields of view and image heights, all within the same landscape with some images from similar viewpoints. The visualisations for the smallest application, containing two turbines, had the widest angle of view, which diminished the perception of distance by a factor of more than four. In response to a request for more realistic single frame images from the local objection group, one visualisation consultant explained that if you cropped 39% from the horizontal field of view of the panorama and 29% of its vertical height, then the fields of view would be that of a single frame. While this approach may be possible using computer software, it is clearly beyond human visual capacity, so why not supply the single frame image in the first place?

It is therefore not surprising that the public are completely confused by the visualisations presented in environmental statements. The lack of any prescription in reality means that windfarm consultants have free licence to produce images that may be misleading, which cannot be verified by any competent authority or understood by the general public. There is, however, a much more practical reason why single frame images must be used: accurate computer–camera matching can only be achieved on a technically correct single frame photograph. If the extent of the windfarm exceeds the field of view of a 50mm lens, each frame has to be montaged separately before a decision is made as to whether the image should be presented in panoramic format using planar (flat) projection or a cylindrical (curved) projection, depending on the audience and purpose of the graphic. If single frame images have been created in the first place, both forms of projection can be created almost at the touch of a button.

For images presented in environmental statements, only planar projection should be used because it does not display the inherent cylindrical distortion and horizontal compression when viewed as a flat image, which is the way the images are invariably viewed. Planar projection can be accurately used for a horizontal field of view of up to 78°, which is much greater than what we see within our detailed vision at any given time. Cylindrical projection, however, has its uses and can be accurately and effectively used in a computerised single frame viewer where wider fields of view are necessary. This is explained in more detail in the next chapter. There is no suggestion that single frame images should replace the imagery or photographic aids for professional assessment; they should be complementary because they serve different purposes and audiences. They also have the advantage of being viewed normally with both eyes when held at a comfortable distance and can be accurately viewed on a computer screen or when projected to a planning committee, so the images are not open to misuse by applicants or misinterpretation by the viewer.

Single frame images are still the most economic, convenient and practical format for on-site assessment for both the public and non-landscape professionals. Animations, 3D imaging and single frame cylindrical viewers may provide a more realistic and comprehensive impression of potential visual impact, but these techniques are more suitable for desktop assessment as they require the use of computer equipment. The Highland Council, who now stipulate additional single frame images, have received unanimous support from community councils, the general public and their own councillors, and for the first time in 16 years, their public and decision makers have images they can easily understand. Appendix 3 contains a brief analysis of the main arguments which are typically put forward against the use of single frame images.

THE VERIFICATION OF IMAGES

Although the Landscape Institute Advice Note states that all the visualisations should be capable of verification by others, images can only be verified if there is a base standard by which all other parameters are measured. Currently this is not possible. The main reason for this is the method used to create the visualisations in the first place. Most panoramic images are formed by cylindrical projection where the turbines are placed on a wireframe model and then superimposed onto a photographic backdrop of overlapping single frame photographs stitched together in third-party software. The photography is usually taken on a tripod fitted with a panoramic head. It is a relatively simple way of producing images because by taking a wide sweep of the landscape in terms of overlapping photographs, the photographer knows that the windfarm will lie somewhere within the overall angle of view.

The method, however, is prone to serious inaccuracies. The stitching of the background photographs is often poorly executed using free software, which results in mismatches in the photographic joins, and there are also cases where the windfarm lies on the join between two planar photographs which have different perspectives. The overall image is also distorted when viewed as a flat image due to curvilinear distortion and horizontal compression and this is further amplified by the use of a 50mm camera lens which compresses the landscape scale due to linear perspective. In most cases, the images have been taken with an SLR camera which is not 35mm format because they have reduced-size sensors, so the original photographs are not technically correct 50mm focal lengths. Although additional single frame images which claim to have a focal length of 50mm have been provided with recent windfarm applications in the Highlands of Scotland, they still cannot be verified as they are 3 × 2 images extracted from the wider panoramas that have cylindrical distortion. More importantly, the images will not contain any metadata because the images have already been processed by stitching software.

This creates a further set of problems. Because the focal lengths most suitable for visual impact assessment are not available as conventional fixed lenses, they can only be accurately recalibrated from a technically correct 50mm image. As the panoramas also have to be vertically cropped because of the way the images are stitched together in cylindrical projection, the focal length is again changed, so we end up with an image which cannot be verified. To enable verification, therefore, it is necessary to have three fixed standards: camera format, focal lengths and image sizes.

THE THREE FIXED STANDARDS

1. The camera format

The 35mm camera is a long-established format. It is also the one recommended by the SNH Guidance and the choice of the windfarm industry. The change from traditional film to digital has not caused any problems, nor should it be an issue. By simply replacing film with an electronic sensor of the same size, the focal length characteristics of the camera lenses have been retained, with the added advantages of much higher resolutions, greatly improved colour accuracy and the ability to verify the images by reading the embedded metadata. Only full-size sensors in excess of 16 megapixels should be used, with all photographic images taken in the uncompressed RAW format only.

The confusion caused by the current range of non-35mm digital cameras should be avoided because reduced-sensor SLR cameras are not 35mm format. However, with the major manufacturers now making a commitment to developing full-frame sensors for their higher-end SLR cameras, the 35mm format will remain the format of choice in the future.

2. The reference 50mm focal length

The 10 reasons for selecting a 50mm fixed lens in the 35mm format as the verifiable photographic base are:

1. The diagonal field of view of around 46° is similar to the field of view of the cones in the human eye which enable us to see detail and colour and to recognise objects.

2. It has minimal distortion.

3. It meets the requirements of the Landscape Institute Advice Note (2011) that the visual representations should 'be based on a replicable, transparent and structured process, so that the accuracy of the representation can be verified, and trust established'.

4. It is the lens recommended by the windfarm industry and all relevant guidance.

5. It enables local authorities to verify the accuracy of the images from the embedded camera metadata.

6. The principle of Leonardo's Window can be applied to an A3 transparency for empirical testing in the field by planning officers (see Fig. 128).

7. Increased focal lengths which are not available as standard fixed lenses can be easily and accurately recalibrated from a 50mm image (see Fig. 129).

8. Because a panoramic image has the horizontal field of view of a wide-angle lens, it meets the requirement of the Landscape Institute's *Guidelines for Landscape and Visual Impact Assessment* (2002) that 'If a practitioner wishes to use an alternative focal length, then a 50mm photograph of the same view should be provided for comparison.'

9. It forms the verifiable base on which images with a wider field of view can be produced using planar or cylindrical projection, depending on the purpose of the photomontage.

10. It is the most economic and readily available fixed lens.

3. The image size

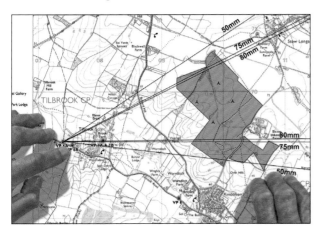

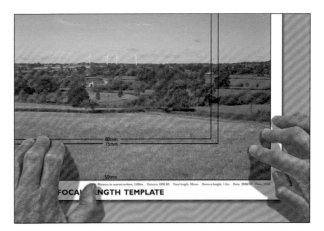

Fig. 125.   Fig. 126.

Standardisation of image size is important because it enables planning authorities to check the accuracy of what is presented. If the base photograph is a verifiable 50mm image of a fixed image size, turbine heights and alternative focal lengths can be easily and accurately assessed by using clear acetate overlays. The Highland Council intends to increase the single frame size from 360mm × 240mm to 390mm × 260mm on an A3 page in line with the image sizes tested in the University of Stirling's focal length perception study (2012). The reason for this is to reduce the amount of white border around the image with the added advantage that the same A3 page layout can be directly downsized to A4 size on a good-quality photocopier without any further modification for the non-technical summaries.

Fig. 127. A3 page layout with an image size of 390 × 260mm with technical information below.

Fig. 128. Applying the principle of Leonardo's Window with one eye, an A3 transparency with a focal length of 50mm is a useful and accurate tool for planning officers and inquiry inspectors/reporters because the real distance cues can be correctly scaled onto the retina (see page 92).

Fig. 129. An A3 printed image with a recalibrated focal length of 75mm viewed at a comfortable arm's length with both eyes is a more practical solution for the wider audience and planning committees. As all guidance states that the visualisations should be assessed in the field, a panorama showing the wider context is not necessary. The image should be held so that it obscures any foreground detail which is not within the vertical field of view of the image.

10

IMAGES SUITABLE FOR ASSESSMENT

Professor Benson in the University of Newcastle Report (2002) recognised that 'Visual Impact Assessment is an integral but distinct part of Landscape and Visual Assessment, including Landscape Character Assessment, Landscape Sensitivity and Landscape Significance.' He also observed that 'the related but distinct area of Visual Assessment is as much a matter for people as it is for professionals' and that 'if viewpoints are to be used as part of any Landscape Assessment, this should be clearly distinguished from Visual Assessment'.

The handover notes from the University following Professor Benson's death in 2004 also recognised that there was 'maybe a need to distinguish between photomontages for context (which may be wide angle and therefore difficult to print for correct viewing distances) and those which are intended to be "realistic" and represent reality/impact which should be printed at the correct size for comfortable viewing'. I totally support this approach, as it is clear that one single image cannot meet the currently defined requirements of professional landscape assessment and the much wider audience within the planning system.

A dividing line needs to be drawn between the professional business of preparing environmental impact statements and the requirements of the planning system itself. The visualisations should be divided into two distinct groups: those for landscape assessment and those for visual assessment, and if combined into a single document, the purpose of the photomontages should be clearly distinguished from each other. This makes suitable images more accessible and affordable to the public and more manageable in the field. Above all, it will cut down considerably on the amount of paper generated because the visualisations can be tailored to their intended audience.

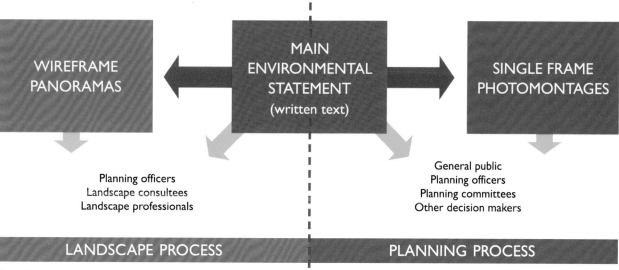

Fig. 130.

METHOD 1: WHERE THE WINDFARM CAN BE CONTAINED WITHIN THE FIELD OF VIEW OF A 50mm LENS

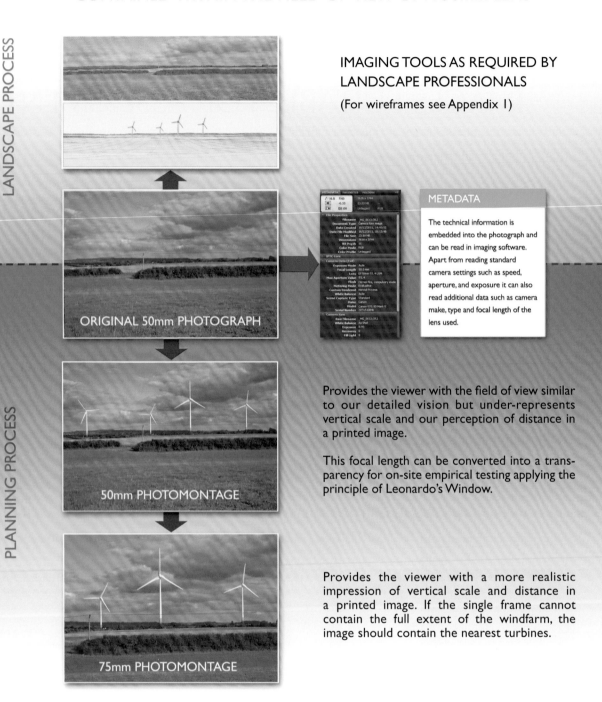

LANDSCAPE PROCESS

PLANNING PROCESS

ORIGINAL 50mm PHOTOGRAPH

50mm PHOTOMONTAGE

75mm PHOTOMONTAGE

IMAGING TOOLS AS REQUIRED BY LANDSCAPE PROFESSIONALS

(For wireframes see Appendix 1)

METADATA

The technical information is embedded into the photograph and can be read in imaging software. Apart from reading standard camera settings such as speed, aperture, and exposure it can also read additional data such as camera make, type and focal length of the lens used.

Provides the viewer with the field of view similar to our detailed vision but under-represents vertical scale and our perception of distance in a printed image.

This focal length can be converted into a transparency for on-site empirical testing applying the principle of Leonardo's Window.

Provides the viewer with a more realistic impression of vertical scale and distance in a printed image. If the single frame cannot contain the full extent of the windfarm, the image should contain the nearest turbines.

Fig. 133. If the windfarm can be contained within the field of view of a 75mm lens, the 50mm image can be omitted at the discretion of the planning authority. If the extent of the windfarm exceeds that of a 75mm focal length but can be contained within the field of view of a 50mm lens, the 50mm photomontage should also be submitted for reference. Framing the nearest turbines may involve a change of camera direction to retain correct perspective.

METHOD 2: WHERE THE WINDFARM CANNOT
BE CONTAINED WITHIN THE FIELD OF VIEW OF A 50mm LENS

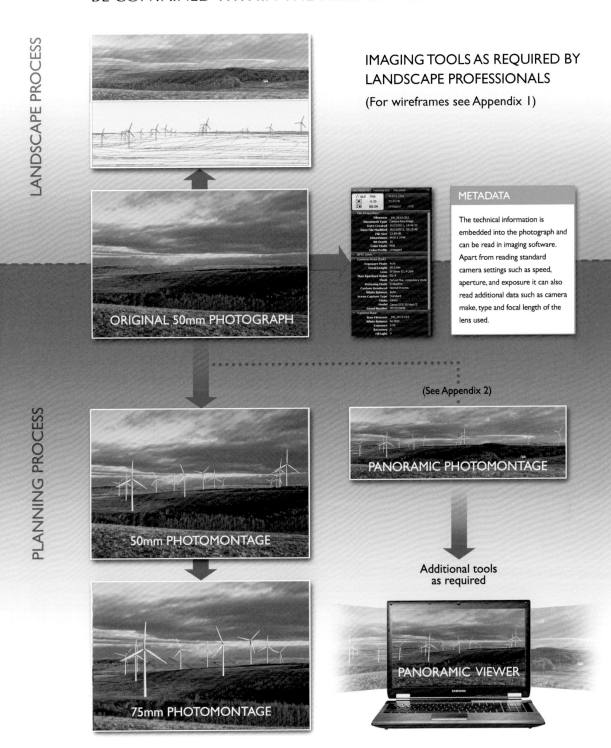

IMAGING TOOLS AS REQUIRED BY LANDSCAPE PROFESSIONALS

(For wireframes see Appendix 1)

METADATA

The technical information is embedded into the photograph and can be read in imaging software. Apart from reading standard camera settings such as speed, aperture, and exposure it can also read additional data such as camera make, type and focal length of the lens used.

LANDSCAPE PROCESS

PLANNING PROCESS

ORIGINAL 50mm PHOTOGRAPH

(See Appendix 2)

PANORAMIC PHOTOMONTAGE

50mm PHOTOMONTAGE

Additional tools as required

75mm PHOTOMONTAGE

PANORAMIC VIEWER

Fig. 134. If the extent of the windfarm exceeds the field of view of a 50mm lens, the verifiable single frame 50mm image must be in the exact centre of the frame, otherwise perspective distortion will occur. Framing the nearest turbines in the single frame images may involve a change of camera direction to retain correct perspective. The panoramic viewer is particularly useful for wide angles of view or cumulative impact assessment because the focal length can be accurately controlled by computer.

EXAMPLES OF METHOD 1

Fig. 135. 50mm single frame image. Turbine height 120m. Distance to nearest turbine 3.85km.

Fig. 136. 75mm single frame image.

EXAMPLES OF METHOD 1

Fig. 137. 50mm single frame image. Turbine height 120m. Distance to nearest turbine 1.18km.

Fig. 138. 75mm single frame image.

EXAMPLES OF METHOD 2

Fig. 139. 50mm panoramic image showing the full extent of the windfarm contained within the horizontal field of view of a 35mm lens (54.4º).   Turbine height: 149.5m. Distance to nearest turbine: 1.5km.

Fig. 140. 75mm single frame image containing the nearest turbines.

EXAMPLE OF METHOD 2

Fig. 141. A slightly different use of a wider 50mm panoramic image. The 28mm horizontal field of view (65.5°) was selected to show the scale of the windfarm development in relation to the Loch which is an amenity area. Turbine height 149.5m. Distance to nearest turbine: 2.1km.

Fig. 142. The 75mm single frame image containing the nearest turbines. The framing required a change of camera direction to retain correct perspective.

ADDITIONAL TOOLS AND APPLICATIONS

Single frame viewers

Single frame viewers should be used for viewpoints where the extent of the windfarm cannot be contained within the field of view of a 50mm focal length and are particularly suitable for cumulative impact assessment and offshore developments. Their main advantage is that a panorama can be correctly assessed for visual impact regardless of the overall field of view and the effect is similar to turning one's head to scan for detail. Single frame images which make up a wider panorama can be quickly converted to cylindrical projection for use in a single frame viewer where the field of view can be extended to a full 360° circle if necessary. The 75mm focal length is precisely controlled by computer so our perception of scale and distance can be maintained throughout.

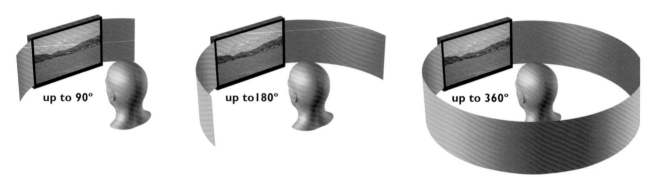

Fig. 143. In a panoramic viewer, any horizontal field of view up to a full circle of 360° can be accurately viewed because the focal length of the camera lens is controlled by the computer software.

Contrary to claims from the windfarm industry that the field of view of the single frame can over-emphasise visual impact because it does not allow for peripheral vision, it has in fact the opposite effect: it will always underestimate. In Chapter 3 it was explained that we have developed our highly sensitive peripheral vision through evolution to detect any motion which could potentially be a threat. In reality, if we were to view the built windfarm, we would be aware of the turbine movement beyond our clear area of vision to the edge of our peripheral field, which would considerably increase their visual presence. However, outside a controlled studio environment where we can view a large curved screen from a considerable distance, it is the most practical solution within the limitations of photography, a printed image, a computer monitor or a projection screen.

Fig. 144. When viewed on a computer screen, panoramic images seriously underestimate our perception of distance.

Fig. 145. In a single frame viewing system, the panoramic image can be panned and viewed at an appropriate focal length.

Fig. 146. If a panoramic image is shown to a planning committee, it should be made clear by the case officer that the image is not representative of scale and distance.

Fig. 147. The same view should also be shown in a single frame viewer where the overall field of view can be panned similar to turning one's head in reality.

A viewing distance is not required because the focal length of the camera lens is used to give a realistic impression of scale. This can be easily recalibrated from a 50mm cylindrical panorama covering the required field of view. In the case of cumulative assessments involving two or more windfarms, compared to the present technique of producing a series of small, wide wireframes in printed form which can seriously underestimate our perception of distance, this viewing system has the advantage of being produced much more efficiently and accurately without the possibility of misinterpretation and can also be recorded on disc or downloaded from websites for screen viewing and economical distribution. As panoramic images can only be viewed on a computer screen, it is important that an on-site assessment is first made with the single frame photomontage taken with a 75mm lens for comparative scaling reference and suitability.

The use of single frame panoramic viewers is particularly suitable for offshore windfarm developments. Although the visualisation requirements are no different from those for onshore developments, there are two important additional considerations. Most coastal communities are orientated towards the sea and distances are very difficult to assess because our seascape is generally devoid of any familiar perspective cues: all we have is a reference foreground and a horizon line.

Most offshore developments also contain a large number of turbines covering a substantial area which cannot always be contained within a 50mm single frame photograph. The 2004 Symonds Report, *Visual Perception versus Photomontage*, which was a post-construction study of the North Hoyle Offshore Wind Farm, and the derivative DTI/DBERR *Guidance on the Assessment of the Impact of Offshore Wind Farms: Seascape and Visual Impact Report* (DTI 2005), make some very constructive suggestions. They recommended a 70 or 80mm focal length. My initial findings are that an 80mm focal length provides the most realistic vertical relationship between any land reference and the horizon line; however, it is an area which now requires more extensive research.

Fig. 148.

50mm transparencies

Using the principles of Leonardo's Window, 50mm transparencies are a useful tool to be used in the field by planning officers, because the image can be viewed with one eye within the context of the wider landscape, where the distance cues are clearly visible within the image itself. It should be noted, however, that there are some practical limitations for use by a wider audience, because the image can only be correctly viewed from the exact height, location and direction of the original camera lens, so it requires very precise placing and location marking. It should also be borne in mind that the centre line of the camera lens is slightly lower than the camera viewfinder at eye level.

Fig. 149. An example of a transparency with the sky removed. When held at the correct viewing distance and viewed with one eye, the image will fit the real landscape cues.

If a 50mm image with an image size of 360mm × 240mm or 390mm × 260mm is converted to a transparency, the viewing distances are 500mm and 542mm respectively. Providing the focal length is verified by metadata, measuring the distance between the eye and the image is unnecessary because the image is simply held at a distance until it fits the real landscape cues. As this can only be accurately viewed in a rigid transparent holder which can be held by the tips of the fingers if necessary, the image can still be viewed at a comfortable arm's length. (Professor Benson observed that images could be viewed up to distances of 600mm when hand-held.)

The Highland Council now require transparencies for selected viewpoints for testing the accuracy of images in the field. An explanation of how they should be constructed is included in their visualisation standards (*Visualisation Standards for Wind Energy Developments*, The Highland Council, 2010). At recent public inquiries, inspectors and reporters have found single frame transparencies to be a particularly useful aid in the field.

Non-technical summaries

The A4 size Non-technical summaries (NTS) presently available free of charge to the public should contain single frame images of the main selected viewpoints and viewpoint maps as agreed with the local authority planning department in consultation with the local community or parish councils. The A4 images are a direct reduction from the 390mm × 260mm images on the A3 pages without viewing information, so very little further work other than printing costs are involved. If the field of view of the overall windfarm exceeds that of a single 50mm image, a panorama containing the full extent of the windfarm should also be supplied for reference only.

Fig. 150. The A4 document should contain detailed location plans and single frame images of selected views.

Animation

Fig. 151.

Animation is a very realistic way of portraying how the movement of wind turbines attracts our attention because the rods in our peripheral vision are extremely sensitive to movement. In the above montage, which illustrates the windfarm development close to a residential cluster, only three turbines are readily identified, but when animated, eight become visible and their movement can be clearly detected through the bare trees.

SUMMARY OF RECOMMENDATIONS FOR PLANNING

CAMERA

Only full frame 35mm format cameras with a fixed 50mm lens mounted on a professional tripod should be used. If the field of view of a windfarm exceeds the field of view of a 50mm lens, a panoramic head should be used if the images are to be stitched together in professional software. The original 50mm RAW files with embedded metadata should be submitted to the planning authority for verification.

Prescriptive

FOCAL LENGTHS

Apart from the 50mm reference photograph, the following additional focal lengths should be supplied: 75mm for onshore and 80mm for offshore windfarms.

Prescriptive

PHOTOGRAPHY AND PHOTOMONTAGE

Photographs should be taken in clear, sunny conditions and free of any unnecessary foreground objects or screening features. Photomontages should be as realistic as possible, taking into account the lighting conditions and sun direction of the photographic background, with the turbine blades unsynchronised facing the camera in a worst-case scenario.

IMAGE SIZES

Images for environmental statements should be printed at 390mm × 260mm regardless of focal length on an A3 page with a small technical information strip, as illustrated on page 81. The same image can be directly downsized to A4 format for non-technical summaries.

Prescriptive

VIEWING INSTRUCTIONS

A viewing distance only applies to panoramic images required for landscape professionals. For A3 single frame images viewed normally with both eyes, an exact viewing distance is not required. The technical information on the page should simply state that 'This image should be viewed at a comfortable arm's length with both eyes (approximately 500mm)'.

A3 ENVIRONMENTAL STATEMENTS

All visualisations should be incorporated in a separate document containing (1) an overall location map showing the viewpoint locations and (2) detailed location maps for each viewpoint along with single frame photographic montages at focal lengths of 50mm and 75mm or 80mm as appropriate. If the field of view cannot be contained within the field of view of a 50mm single frame, a planar panoramic image containing the extent of the windfarm only should be included for reference.

A4 NON-TECHNICAL SUMMARIES

These documents provided free to the public should contain reduced-size versions of the A3 single frames at all the required focal lengths. The selected viewpoints should be agreed with the planning authority. Where a windfarm cannot be contained within the field of view of a 50mm single frame, an additional panoramic image showing the full extent of the windfarm only should be provided for reference. No viewing instructions are required.

SINGLE FRAME VIEWERS

Where the field of view of a windfarm or windfarms cannot be contained within a 50mm single frame image, the panorama should be incorporated into a viewer for accurate visual impact assessment. Images should be formed using cylindrical projection.

11

CONCLUSION

The recommendations and observations of John Benson (University of Newcastle, 2002) were very important in identifying the main problems with the presentation of visualisations in windfarm applications. Despite Scottish Natural Heritage's claims that the SNH Guidance (2006) was built upon his initial findings, the recommendations regarding the public role in visual impact assessment were overlooked because the independence of academia was compromised by a small group of consultants. The technical sections of the Guidance were never subjected to any peer review, no post-construction studies of existing windfarms have been initiated since 2002, nor was the methodology ever subjected to any field tests before its publication.

The lack of any real academic authority which can encompass all the disciplines of cognitive science, photography and computer visualisation has therefore left the field wide open to highly questionable invention and manipulation. The problems identified by Benson and in many of the research studies noted in Appendix 6 have been largely ignored and even the bizarre impracticality of the viewing methodology has surprisingly raised little concern. Over a decade, it has all been gathered up into terms like 'industry standard' or 'best practice' to counter any complaints from planners and the general public. SNH and the landscape profession have neglected to give appropriate weight to the needs of the different audiences within the planning system and the different visual requirements for landscape assessment and visual impact assessment. In photographic terms, the horizontal field of view is the priority for landscape assessment; the vertical field of view is the priority for visual impact assessment.

Most environmental statements are now produced in their entirety by large landscape and environmental management practices who embrace expertise in almost all the specialist areas required in EIA production, including visualisations. These practices, largely representing clients in the wind energy sector, are reluctant to question current formats or accept changes that contradict current 'best practice' because this would not merely raise issues of inconsistency, it could also leave a practice vulnerable to challenge on any of their projects. This process-led and all-embracing approach to EIA production has inevitably removed most of the independent expertise and objectivity which should maintain a balance of opinion. This is particularly the case in the vital area of visual representation, which has become increasingly formulaic and dependent upon proprietary windfarm software packages. I do not believe that it was anticipated that the European EIA Directive would spawn an entire in-house industry, running totally counter to the objectivity implicit within the EIA process, nor could anyone have anticipated the change in status and power it has bestowed on the landscape profession and their resulting control over the form of all visual material submitted to the planning process relating to wind energy projects. All of these outcomes may be typical of the law of unintended consequences; however, for our planning system and the trust placed in the objectivity of the EIA process, the consequences have been significant.

Complexity is regularly used as a defensive bastion to deter criticism and to confound government inspectors/reporters, planners and politicians who, mainly, through lack of knowledge on the subject, immediately retreat at the first mention of mathematical formulae, the science of human vision and characteristics of photographic lenses. As a result, misleading statements have been made by landscape witnesses at public inquiries regarding the importance and 'correct' application of viewing distances. Improved visualisations are often conjured up at this final stage in the planning process, when they should have been provided to the public during the much earlier consultation period; counter-visualisations commissioned by third-party objection groups and communities, involving single frames and focal lengths greater than 50mm, are regularly disputed because they do not conform to 'best practice'. Countless hours are also wasted at public inquiries debating the importance of context, viewing distances and representative focal lengths using well-rehearsed routines by applicants' legal teams.

These confusions and contradictions only serve to engender a growing public distrust of current practice. The multidisciplinary Beyond Nimbyism project, involving several British universities, led by Professor Patrick Devine-Wright (*Beyond Nimbyism Project Report*, University of Manchester, 2005–09) sought to 'deepen understanding of the factors underlying public support and opposition to renewable energy technologies'. Over 3,000 people participated in the project, which consistently found a lack of trust in developers and strong concerns about the fairness of planning procedures. In discussion with Professor Devine-Wright (personal comment, 2010), he confirmed that the subject of the truthfulness and accuracy of photomontages was one of the issues regularly raised by the participants who took part in face-to-face interviews and focus groups.

While it is the stated intention of most governments to promote wind energy development, the recorded, perceived lack of integrity of developers and their consultants can only serve to obstruct these ends and subvert a policy which requires considered and careful planning. The Landscape Institute, which should strive for the best possible professional technical standards to stand alongside the accuracy and transparency it promotes in its criteria, has failed to stipulate any defined standards and simply supports the SNH Guidance (2006) which members are strongly advised to follow. Regardless of what guidance is quoted in environmental statements, it is also the landscape consultant's professional duty to ensure that the windfarm visualisations meet the requirement of the European EIA Directive that the impact should be properly understood by the public and the competent authority. Clearly this is currently not the case. Although it is argued by landscape consultants that reporters and inspectors can make their own judgements in the field using the panoramic photomontages as visual aids, they do not have a fair comparison on which to make that assessment. The human brain is not capable of defining the small 3 × 2 photograph which forms part of a much wider panorama, mentally enlarging it to full page size, while at the same time zooming in to the image to increase its focal length until it matches the vertical scale of the real landscape. It is simply not possible and it is one of the main reasons why the visualisations continue to be the subject of such widespread scepticism.

Creating images for visual impact assessment is not the exclusive domain of landscape consultants; it also applies to other disciplines, such as engineering and architecture, and is a highly specialist subject. The purpose is to convey, through the use of imagery, a clear and realistic understanding of potential visual impact without the possibility of any misinterpretation. Members of the public, particularly local people who know their landscape and may even be directly affected by the development, have as much right to make an informed judgement on potential visual impact as a landscape architect. To enable them to make that assessment, they have to be provided with accessible and realistic images which are not open to misinterpretation and can be easily understood.

Because the present visualisation methodology can seriously under-represent the visual impact of windfarm developments by anything up to a factor of four and sometimes more, it is not possible for anyone to accurately predict the cumulative effect of the many applications presently in the UK planning system. This is now a matter of concern to many councils because our landscape does not stop at local authority borders; it is a continuous flow of space until it reaches our shores, where the visual impact of such structures now extends beyond our coastlines with the construction of large offshore windfarms. Good mitigation is also inhibited by the lack of reliable or functional visualisations which can help planners to contain intrusion by selectively removing or repositioning visible turbines.

Visual 'accidents' are already happening in the Highlands of Scotland, with developments impacting on popular tourist views which were never envisaged when the schemes were approved. The distant visibility of windfarm developments is also much greater than that predicted by applicants' landscape experts: under certain lighting and atmospheric conditions, turbines at a distance of over 40km can be clearly visible to the naked eye.

The field of cognitive science is a very complex one. There is no exact scientific formula for producing photographs which best represent what we see in a landscape involving considerable distances, as the University of Stirling study (Hunter and Livingstone, 2012) has highlighted. While a 50mm transparency can be empirically tested in the field, applying the principle of Leonardo's Window, there is overwhelming evidence that a printed image with increased focal lengths provides all audiences with a more realistic representation of the scale and proximity of wind turbines.

The *Guidelines on the Environmental Impacts of Windfarms and Small Scale Hydroelectric Schemes* (SNH 2001) recognised that 'although a 50mm camera lens is commonly accepted to best represent the naked eye, selective focusing by a viewer, particularly to a prominent focus such as a windfarm, may mean that a telephoto lens of around 80mm is more truly representative'. This has also been the conclusion of a number of councils, independent studies, photographers and members of the public for many years. Over ten years, we have lost an irrecoverable opportunity to create visualisation standards which are functional, accessible and suitable for a democratic planning process and we have simply come full circle. The power of prediction which computer visualisation technology offers us and the opportunity to selectively mitigate visual intrusion have been squandered, justified by the application of a viewing method which has no practical or technical credibility, developed behind closed doors and of dubious parentage.

It is accepted that photography has limitations and that a two-dimensional printed image can never precisely represent human vision, but we are aware of this through experience and mentally adjust for it. Photomontage can nevertheless be used as a much more accurate prediction tool within the planning system than is presently the case. The quickening pace of innovation in visual media will create many new techniques which will inevitably become the next panacea for realistic representation of future developments. Some may have a contribution to make but I remain sceptical that any new visual presentation technique can ever replace a simple straightforward photograph as the most familiar, accessible, and practical base for computer visualisations.

In the last decade, increasing access to computer simulation techniques for wind energy developers and their consultants has also led to an imbalance between the photography and the computer-based montaging process. The base photography is too often detached and sub-contracted to a third party, consequently the visualisation specialist is removed from the cognitive experience of the landscape and the lighting conditions for accurate image adjustment. The original photograph must be the heart of any specialised photomontage procedure providing the canvas for realistic imaging and more importantly, the parameters for technical verification.

Fixed photographic and image sizes are therefore vital foundations for the basic standards I have outlined. My suggestions are purposely simple in order to define clear photographic and image presentation parameters which can be easily attained, tested and verified, and from which other presentation solutions can be developed. I have also tried to take full account of future viewing techniques, practicality and accessibility for a universal audience. Both experience and analysis suggest that viewing distances should be excluded from the planning system for the wider audience, for whom the only permissible reference would be a comfortable arm's length (approximately 500mm). What is comfortable for the viewer should always dictate the technical data and not vice versa.

For professional assessment I am sure that landscape architects will make up their own minds over the visual tools they require and there are many opportunities for improvement. However, for the planning system, prescription is long overdue. For a system which is regulated at so many levels, it is extraordinary that we have shied away from regulating visualisation for planning. It is clearly apparent that self-regulation does not work and should therefore not even be considered.

It has also become evident that there is a widespread lack of knowledge of photography, its uses and implications among landscape professionals, which suggests that their Institute should make this important subject a fundamental and essential part of the education curriculum. Council planning officers also have to take a much more active role in ensuring that visuals submitted with applications are accurate and verifiable representations of the visual impact, and that they are not open to misinterpretation and are appropriate for public consultation. If this is not the case, they should use their enforcement powers as the competent authority to obtain improved visualisations which can be verified and more clearly understood.

It is recognised that this book is being written when change is in the air over future visualisation requirements. The SNH Guidance is under substantial review, the Landscape Institute's 'Blue Book' is being revised and the Highland Council's Visualisation Standards are to be updated in the light of feedback and research over the last two years. It is not possible to predict what will emerge, but it is to be hoped that some of what is necessary will prevail. As well as establishing fixed and verifiable foundations upon which to build, all new guidance should also urgently address the need for a code of ethics for landscape visualisation as proposed by Professor Stephen Sheppard.

Basic consideration of the science involved suggests that the viewing methodology as exemplified by the current practice amounts to visual deception. It may be correct in terms of perspective geometry, but this has resulted in an almost comical method of looking at images which is unnatural, impractical, unreliable, rarely applied and fails to provide a representative image. As a matter of priority this must be corrected. The EIA photomontages presently submitted under-represent the perceived scale of the landscape and the wind turbines, overestimate the distance to the development, overestimate our field of clear vision and compress and distort the overall image to a considerable degree.

Planning visualisation is all about predicting what we will actually see if a development is constructed. This means that our starting point can only be to match what the human visual system perceives in terms of scale and find the best way of presenting reliable and practical solutions through the straightforward use of photographic images.

Perspective or perception? I hope the reader is now in a position to decide.

For further information, please refer to www.windfarmvisualisation.com

Fig. 152. Remember Hong Kong? This image contains three of its most famous buildings, which are not relative in scale because they are at different distances. The diminishing effect is due to linear perspective.

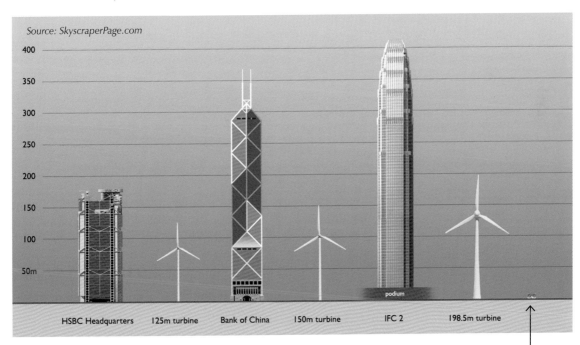

Fig. 153. The true comparative heights of the above three buildings along with turbines to scale at 125m, 150m and 198.5m (the tallest turbine under manufacture at the time of publication). It can be seen that the tallest turbine is almost half the height of Hong Kong Island's highest skyscraper. The scale of a typical rural house in Scotland is added for comparison.

APPENDICES

APPENDIX 1

TECHNICAL ASPECTS OF WIREFRAMES

Wireframes used as an aid to on-site assessment by landscape professionals are made up of square grids or triangular grids depending on the software used. As a result, straight lines are introduced into an otherwise organic landform which has the exaggerated perspective of a wide-angle lens. Because they are usually formed by cylindrical projection, they also have the same curvature distortion and horizontal compression inherent in photomontages if viewed as a flat image. Although designed for professional use, wireframes are increasingly submitted to the wider audience, particularly at public local inquiries. The public find them confusing and difficult to orientate because there are no familiar landscape features. At a recent inquiry in the Highlands of Scotland, a member of the public who was complaining about misleading visualisations compared them to images containing small insects stuck in a spider's web.

I would therefore recommend the use of wirelines or solid terrain modelling, as shown in the two images below, formed by planar projection, which removes the problem of geometric perspective lines and image distortion while maintaining an accurate landscape profile. However, this would be a matter for landscape professionals to decide and is not the focus of this book. Different combinations of wirelines and photographs can, however, be effectively used in single frame images to inform a wider audience, depending on the purpose of the graphic. Figures 156–159 show some examples of appropriate applications.

Fig. 154. A solid terrain model using planar projection removes the problem of geometric perspective lines and maintains an accurate landscape profile.

Fig. 155. As an alternative, wirelines are useful where landscape features are visible. They can easily be converted into transparencies for empirical testing.

Fig. 156. Photomontage showing the predicted view of the windfarm on the setting of a Grade I listed building. Turbine height: 125m. Distance: 2.03km. Focal length: 50mm.

Fig. 157. A wireline overlay showing the location of all the turbines, which will appear or disappear according to our viewing position.

Fig. 158. A wireline overlay showing the location of the turbines photographed from the entrance gateway to the Grade I church and churchyard. As you move along the public footpath out of frame to the left, the same turbines appear on the other side of the church due to motion parallax.

Fig. 159. The location of the same turbines from the other side of the church. As both images have a focal length of 50mm, the photomontage can be converted into a transparency. (See page 92.)

APPENDIX 2

TECHNICAL ASPECTS OF PANORAMAS
50mm focal length

The panoramic presentation format should be used where the full extent of a windfarm cannot be contained within the overall field of view of a 50mm image and can be created by using planar or cylindrical projection. When viewed as a flat image, planar projection does not demonstrate the distortion or horizontal compression inherent in cylindrical projection, so the accuracy of the overall field of view and the landscape profile can be maintained without cropping or distortion. Planar images are also easier to use because they do not need to be curved in an exact arc, which makes them more manageable on site, since the A3 document can be tensioned to prevent wind distortion.

Most importantly, the three fields of view which define the 50mm focal length are maintained and can be verified by metadata. For the projection to be accurate, the single frame photograph containing the windfarm or the centre of the windfarm must be exactly central in the panoramic frame, otherwise perspective distortion will occur. This requires the photographer to accurately frame the direction of the camera lens by using either compass bearings or computer renders of the development with known reference points.

Fig. 160. A panoramic image created in planar projection with the horizontal field of view of a 28mm wide-angle lens. The 50mm single frame image is defined in the centre of the image.

While I would recommend that panoramic images are used for reference only, if they are to be used scientifically, there are a number of points to be observed regarding their technical aspects. If a viewing distance is used, to meet the current minimum viewing distance requirement of 300mm, a panoramic image has to be 144mm high if the focal length is 50mm. To comfortably fit on an A3 page, allowing for a binding edge and page borders, an image size of 385mm × 144mm will accommodate a 50mm planar panoramic image with a field of view of 65.5°, which is the horizontal field of view of a 28mm wide-angle lens.

Two planar 28mm images side by side spread over two pages in an A3 report would also make up an overall field of view of just over 130°, which is approaching the lower threshold of the visual field of one eye including peripheral vision. A panorama taken with a 24mm lens which has a field of view of 73° would have to be shown on extended A3 pages, which would make them more inaccessible to the public because they can only be printed by a professional output centre.

The extent of the technically correct 50mm photograph should also be defined in the centre of the image. This serves two purposes: it shows the viewer the extent of the panorama which will be contained within our in-focus vision at any given time and it allows the local authority to compare the defined area with the original 3 × 2 single frame photograph submitted with the metadata for verification purposes. Because the image is made up of overlapping photographs, the camera must be mounted on a panoramic head which has to be levelled in both the vertical and horizontal planes to prevent distortion.

Two warnings should be clearly stated on the image page:

1. The image should be viewed with one eye only from a fixed distance of 300mm.
2. Unless the image is converted to a transparency, the vertical scale will appear compressed.

There is, however, a much simpler way to produce a planar panorama. The 50mm single frame image which is centred on the windfarm development can be superimposed into a lens-corrected image taken with a wide-angle lens and cropped to the vertical height of the 50mm image, providing they share exactly the same perspective and point of focus. Providing the camera and tripod are locked in position, it only involves a change of lens. This technique is free of the restrictions of a panoramic head, which cannot effectively work if the elevation of the windfarm is considerably different. An example of this would be a panoramic image taken from a valley floor where the windfarm is located upon an adjacent high ridge. This would also remove the necessity for portrait framing, which changes the focal length of the camera lens.

When we view the world around us, we keep our head vertical so the horizon is always horizontal, but besides constantly scanning the scene for detail, we also tilt our head up or down depending on our point of focus. This is clearly illustrated in Taylor's sketch of Leonardo and his window on page 21. In reality, we see more below our line of sight than we do above the line, so it allows the photographer more flexibility in framing a view similar to what we would see according to our point of focus. As a transparency, it also conforms to the principle of Leonardo's Window when aligned with the real landscape.

Fig. 161. A 75mm panoramic image provides a more realistic impression of landscape scale when viewed normally with both eyes at the same distance of 300mm.

When viewed as a printed image where the distance cues are no longer visible, increasing the focal length to 75mm provides the viewer with a more realistic impression of landscape scale when viewed normally with both eyes at the same viewing distance of 300mm. So that the images can be compared like for like based on the same image size of 385mm × 144mm on an A3 page, a 75mm focal length will provide a horizontal field of view of 46° (to the nearest round number).

APPENDIX 3

COMMON ARGUMENTS AGAINST SINGLE FRAME IMAGES

In most cases, requests for single frame images are resisted by developers and their consultants on the grounds of not conforming to best practice or the SNH Guidance. The most common reasons are summarised below:

A 50mm single frame can be extracted from a cylindrical panoramic image. It is not possible to extract a verifiable 50mm from a panoramic image because they are invariably cropped, having been formed by cylindrical projection. Because cropping changes the focal length as explained on page 36, any 3 × 2 image extracted from the centre of the panorama will not be a technically correct 50mm focal length and will also be horizontally compressed. More importantly, any single frame image extracted from a panorama will not contain any metadata, so the image cannot be verified by the local authority, nor can the image be used as the reference base for checking the accuracy of any recalibrated focal lengths.

Single frame images involve additional work. This argument has no validity because if panoramic images have been accurately constructed in the first place, the 50mm single frames have to be created before the wider images are formed. The 75mm images can also be quickly and easily extracted from the 50mm reference frame using 3D camera software or pre-prepared computer templates, involving little additional cost.

The image does not show enough context. The image is not meant to show the wider landscape context: this is already shown in the wireframe image used by landscape professionals. The single frame images are for the separate assessment of visual impact, where the vertical field of view is used to provide a realistic impression of the scale of the development relative to distance when viewed from the actual viewpoint. Providing the image can be viewed at a comfortable distance which allows the viewer to directly compare the photographic image with the real landscape, there is no need for a wider photographic context: it is already there in front of us.

The image does not specify an exact viewing distance. A viewing distance can only be accurately applied to a transparency when viewed with one eye because the methodology is based on the principle of Leonardo's Window: it is how artists created correct linear perspective to create the illusion of depth. However, we do not view images in this way; we do so with both eyes at a comfortable distance, so a viewing distance cannot apply. Our perception of scale and distance is achieved by using a combination of image size and the focal length of the camera lens, which is in line with the University of Newcastle Report (2002) recommendation that what is comfortable to the viewer should dictate the technical detail, not vice versa.

The 75mm image has to be viewed at beyond arm's length. This is a somewhat misleading statement. On page 23, we explained that if we enlarge the original 36mm × 24mm image taken with a 50mm lens by a factor of ten, the point of projection or viewing distance is 500mm. If the focal length is increased to 75mm, the distance increases to 750mm, which is beyond arm's length. This is only correct if we are viewing a transparency with one eye where the real distance cues can be accurately scaled onto the retina because we are applying the principle of Leonardo's Window. We do not see the two-dimensional image projected onto the retina because it has already been processed by the brain where we take distance into account to recalibrate the scale of more distant objects by size-constancy scaling, as explained in Chapter 4. When we view photomontages in an environmental statement, we view flat printed images, devoid of any distance cues, with both eyes. In photographic terms, by increasing the focal length of the camera to closely match the vertical scale of the landscape, we are simply compensating for size constancy when the image is viewed from the same distance as the 50mm printed image. Again, what is comfortable for the viewer dictates the technical detail.

The public will find too many formats confusing. The SNH Guidance states that it is 'not targeted at the general public given its specialist nature and technical content'. The Highland Council, who are accountable to the public, have found that by providing them with single frame images, they have a much clearer understanding of the potential visual impact. The panoramic images are a requirement for landscape professionals only, but for local people they are unnecessary because their cognitive map of the area is highly developed. Single frame images are the easiest to understand and are also the most practical to produce because they can be printed on standard paper sizes without the need for one-off panoramic prints, which are expensive.

The images overemphasise visual impact. Images within the focal range of 50–80mm cannot overemphasise visual impact because there is no depth information in a two-dimensional photograph and because more distant objects will always appear smaller in relation to foreground objects. Visual impact could only be overemphasised if the focal length of the image is increased so that the vertical scale of the landscape looks greater or closer to what we see in reality. A 105mm lens, which is a popular choice for portrait photography, would, for example, give the perception of being too near. As the images should be viewed in the field, the wider photographic landscape context is not necessary.

It is the field of view that is important, not the focal length. This requires a two-part answer. Firstly, while the field of view may be an important requirement for landscape professionals, it is not readily understood by anyone else. Even the most experienced photographer would have some checking to do if he was just asked to produce a photograph with a 39.6° horizontal field of view, because it is not how photography is specified. Apart from the fact that all guidance states that the focal length must be included in the technical data on the image page, the accuracy of the image can only be verified and the viewing distance calculated if the focal length is known. This can only be confirmed by analysis of the camera metadata.

Secondly, while the field of view may be the priority for landscape context, focal length is the priority for visual impact assessment. In reality, the field of view of a panorama is made up of overlapping photographs stitched together horizontally to make up an overall field of view of a wide-angle or ultra wide-angle lens. Regardless of what focal length is used, the horizontal field of view will always remain the same. A 90° field of view made up of stitched 50mm images, for example, will be no different from a 90° view made up of stitched 75mm images. The only difference will be in the vertical height of the images, as shown in Fig. 97 on page 59. For visual impact assessment, however, focal lengths have an important effect on our perception of scale and perceived distance in a landscape as shown in Figs. 7 and 8 on page 4. This can only be achieved by using single frame images where the three fields of view which define the focal length are technically correct. Single frames require a return to defined focal lengths, which have recently been excluded from technical data in favour of a constructed angle of view.

The images are too prescriptive. Prescription may not be required by landscape professionals, but it is necessary if the images are to be verified as required by their Institute. This can only be achieved by establishing a reference photographic standard which is widely accepted as a 50mm focal length in the 35mm format. Fixed sizes are therefore necessary for single frame images, but there are no such restrictions on the field of view contained within a panorama. If presented on an A3 page, which is the most convenient and practical format for use on site, the field of view cannot much exceed 65.5° if a viewing distance is used. However, if presented in a single frame viewer, which is explained on pages 90 and 91, the field of view can be accurately extended to a full 360° circle. The argument that prescription would be too restrictive because the required visual context of view can vary from project to project therefore has no foundation.

In some cases, the single frame 50mm and 75mm images cannot contain the full extent of the windfarm. In such cases, the single frame photomontages should contain the nearest turbines. A wider panorama should also be submitted containing the full extent of the windfarm, but with the appropriate warnings. By providing a reference panorama of the wider view containing the windfarm, a 50mm single frame image which adequately represents the field of view of our detailed vision, and a 75mm single frame image which is more representative of scale and distance, an informed overview can be made without the possibility of misinterpretation. This is especially important in cases where a consultee may have to rely on desktop assessment, although it does clearly state in the Landscape Institute Advice Note that all visualisations must be assessed in the field. The full extent of the windfarm can also be viewed in a panoramic viewer if considered necessary by the planning authority.

There is not much difference between an image with a 52.5mm focal length and a 50mm focal length. There is a perceptible difference, but that is not the point. The accuracy of the 50mm focal length can be directly verified by the metadata. By having a fixed print size, the accuracy of recalibrated focal lengths and turbine heights can be easily checked by the planning officer using standard acetate templates.

APPENDIX 4

VIEWING DISTANCES AND HORIZONTAL COMPRESSION

Regardless of whether the image is a single frame or a panorama, the viewing distance is the same and can be simply calculated on its height. The two photographs below are similar to the images shown on page 173 of the SNH Guidance and are available in both the printed and the downloadable version. The left image shows a normal single frame image while the right image shows the same image when cylindrical projection is applied.

Fig. 162. Photograph taken with a 50mm lens.

Fig. 163. Photograph transformed to cylindrical projection.

Rephotographed and adapted to replace figures B9 and B10 in Visual Representation of Windfarms: Good Practice Guidance.

If we now look at the vertical centre line of both photographs, the height is exactly the same, but the image formed by cylindrical projection is curved and shrinks towards the edges because we are now viewing a curved surface on a flat page. If these images were now viewed as a transparency in the real landscape using one eye only, the viewing distance at the centre of the photograph would be the same for both images, but the image formed by cylindrical projection on the right would have to be curved towards the eye to fit the real distance cues.

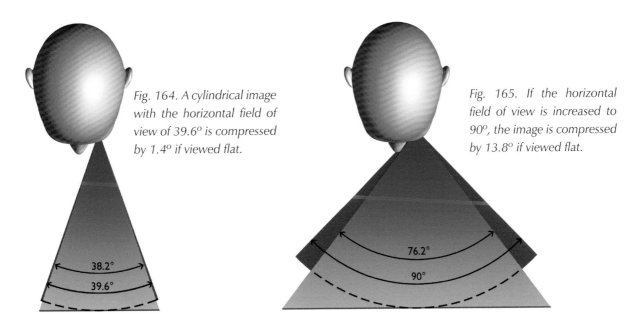

Fig. 164. A cylindrical image with the horizontal field of view of 39.6° is compressed by 1.4° if viewed flat.

Fig. 165. If the horizontal field of view is increased to 90°, the image is compressed by 13.8° if viewed flat.

Because most panoramic images use cylindrical projection, the images are compressed horizontally when printed on a flat page and the degree of compression is progressively increased as the overall field of view becomes wider. A 39.6° field of view is compressed by 1.4° but a 90° field of view is compressed by 13.8°.

Based on an image height of 144mm to meet the minimum viewing distance of 300mm, the chart below shows the degree of horizontal compression for various focal lengths relative to the overall field of view.

	18mm	24mm	35mm	50mm
Horizontal field of view when image is curved	90°	73.7°	54.4°	39.6°
Actual image width on page	471.2mm	385.9mm	284.8mm	207.3mm
Horizontal field of view when image is viewed flat	76.2°	65.4°	50.8°	38.2°
Difference in contained angle	13.8°	8.3°	3.6°	1.4°
Viewing distance	300mm	300mm	300mm	300mm

Fig. 166. The above fields of view have been taken to the nearest single decimal place and image widths rounded down to the nearest millimetre.

To illustrate that the viewing distances are the same for both single frame and panoramic formats, if we now apply the two different SNH formulae referred to in Chapter 6 page 52, the results are exactly the same.

For a single frame image:

$$d = \frac{w}{2 \tan\left(\frac{A}{2}\right)}$$

where d is the correct viewing distance in mm, w is the image width in mm, A is the horizontal field of view in degrees and tan is the trigonometric function.

For a panoramic image:

$$d = \frac{180\,w}{\pi A}$$

where d is the correct viewing distance in mm, w is the image width in mm, A is the horizontal field of view in degrees, and π has its usual geometric value.

For an image with a field of view of 39.6º (50mm focal length), the viewing distance is calculated as follows:

$$d \;=\; \frac{180 \times 207.3}{3.14 \times 39.6} \;=\; \frac{37314}{124.3} \;=\; \textbf{300mm} \text{ (to the nearest round number)}$$

For an image with a field of view of 90º (18mm focal length), the viewing distance is calculated as follows:

$$d \;=\; \frac{180 \times 471.2}{3.14 \times 90} \;=\; \frac{84816}{282.6} \;=\; \textbf{300mm} \text{ (to the nearest round number)}$$

If the above formulas are also applied to the 24mm and 35mm focal lengths, the viewing distance is 300mm.

APPENDIX 5

EFFECT OF FOCAL LENGTH ON PERCEPTIONS OF SCALE AND DEPTH IN LANDSCAPE PHOTOGRAPHS: IMPLICATIONS FOR VISUALISATION STANDARDS FOR WIND ENERGY DEVELOPMENTS

by Peter D Hunter* and Duncan F. Livingstone, University of Stirling.

1. **Background.** The purpose of this study was to investigate the effect of focal length on public perceptions of scale and distance in landscape photographs.

2. **Methods.** We have conducted 362 interviews (to date) with members of the general public at 8 viewpoints in the Central and Highland regions of Scotland. The participants were shown a series of photographs of specific focal points taken at focal lengths of 50, 60, 70, 80, 90, 100 & 110 mm. The images were 390mm × 260mm, mounted on A3 polyboards and presented to participants in random order. The distance to the focal area varied across the viewpoints from 0.18km to 24km. The participants were then asked to specify which focal length, in their opinion, provided the most accurate representation of scale and perceived distance of the focal area in the landscape. We also recorded standard demographic information on the interviewees.

3. **Preliminary results.** We found that less than 5% (n=16) of participants believed that a 50 mm focal length provided the best representation of scale and distance. The histogram in Fig 1a shows the full distribution of focal length preferences across the 362 interviewees for all viewpoints combined. The most common (modal) focal lengths were 70mm and 80mm (n=85; 23.5%). The cumulative density function in Fig 1b shows that the mean of the focal length distribution was 79.0 mm with a standard deviation of 14.9mm.

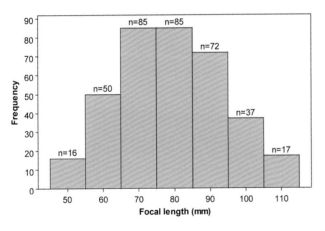 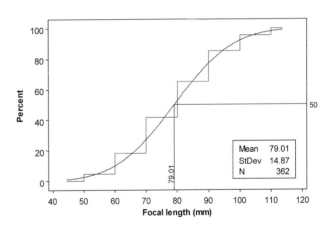

Fig.1 (a) The distribution of focal length preferences; and (b) the cumulative density function and estimated parameters from a Gaussian fit to the observed distribution.

There was marked variability in focal length preference between the various viewpoints used in the survey. The mean focal length preference for viewpoints with >10 responses varied from 73.0mm for the view of Creag Nay Hill from Urquhart Castle to 89.3mm for the view of a residential dwelling adjacent to Temple Pier again from Urquhart Castle. The effect of distance on focal length preference appears to be complex. However, it is significant that the number of respondents selecting the 50mm focal length was never greater than 12.5% for any single viewpoint irrespective of the distance to the focal point.

4. Preliminary conclusions.

- The vast majority of participants were of the opinion that the photographs taken using a 50mm lens underestimated the scale and overestimated the distance to the landscape focal points for distances between 0.18km and 24km.

- This finding suggests that the use of a 50mm single frame image for the visualisation of wind energy developments is inappropriate.

- The majority of respondents selected a focal length greater than 75mm. This finding broadly agrees with the use of focal lengths of 70mm (<1.5km) and 75mm (>1.5 km) as stipulated in the current Highland Council Standards for Wind Energy Developments.

These preliminary results are released courtesy of The Highland Council, Planning and Development Service. The detailed study report is expected to be available in April, 2012. For further information contact Planning & Development Service, The Highland Council, Inverness, Scotland.

*Biological and Environmental Sciences, Faculty of Natural Science, University of Stirling, Stirling FK9 4LA
Email: p.d.hunter@stir.ac.uk Tel: +44 (0)1786 466538

APPENDIX 6

RESEARCH PAPERS, STUDIES AND ADVICE

Part of the brief for the University of Newcastle Report was to identify and assess any relevant work on visibility, visual impact and significance. John Benson unearthed a weight of research and a wide range of guidance and opinion on the detailed issues of distance, visibility and significance for windfarms. On the more particular issue of photography and photomontage visualisation, the result was rather more sparse and sporadic.

In the course of writing this book I have sought out all the research papers referred to by Professor Benson and the authors of the SNH Guidance. Some interesting early papers do exist although some have proved difficult to trace and many are now out of date in the light of current scientific knowledge and changes in photographic technology. I have also traced other relevant published material in the course of my own research and have included all those which have something pertinent to contribute.

The commentary on the papers, studies and advice has been arranged chronologically so that any patterns of emerging opinion can be discerned from the 1980s to the present time.

Steven Shuttleworth provides an early source for the use of wide panoramic photographs by the landscape profession. In his 1980 article **The use of photographs as an environmental presentation medium in landscape studies** (*Journal of Environmental Management* 11) he investigates their use as surrogates for landscape character and quality assessment. He believes that perceptual ambiguity can be reduced if the field of view is as large as possible. However, while he finds such photographic simulations reliable in dealing with the overall perception of landscape, he finds them less reliable when dealing with perception of detailed elements and characteristics in the landscape. Shuttleworth recognises that 'there is a need for empirical research to validate the use of photographs as an environment presentation medium in landscape studies'.

A very specialist paper, **Distance perception as a function of photographic area of view**, 1989, published by **Kraft and Green** in *Perception and Psychophysics* (45) is the only paper traced which examines distance perception of depicted objects as a function of the photographic area of view using five different focal lengths and requiring subjects to make distance judgements. The authors observe that in the natural environment, the visual field begins at the face. With a photograph, however, the scene begins some distance away and the observer makes an adjustment. This adjustment then changes as a function of the photographic area of view. They conclude that 'widening the photographic area of view by shortening the focal length alters the user's perception of distance. Photographic area of view, then, is an important consideration in the formatting of visual displays.' The actual photographic distances of 20m to 320m in this study are rather short for our purposes, possibly because research work at the time in this field was focused towards the needs of airline pilot training and the development of visual imagery for simulators.

The Countryside Commission publication CCP 357, **Wind Energy Developments and the Landscape** (1991) lists in its suggestions a need to 'represent an impression of the likely impact. The most successful form of simulation used to date in relation to proposed windfarms is the colour photomontage and the impact of movement can be conveyed with the assistance of a video montage.'

The **Stevenson and Griffiths** 1994 study, **The Visual Impact of Windfarms: Lessons from the UK Experience** (ETSU W/13/00395/REP) involved a post-construction audit of eight windfarms in England and Wales; each site was visited four times in different seasons and six viewpoints were analysed at each site. The detailed findings of this study are no longer applicable because the turbine heights have now more than doubled and visualisation methodology and technology has changed significantly, but its value is that it was conducted in the field and was based on photographic comparison. Single frame photographs are used along with panoramas with clearly visible joins where the spread of the development requires more than one frame. The report does make one very important observation, that 'where photographs smaller than A3 are used, the turbines on the photomontage appear smaller than in reality'. This is the first reference to the importance of image size. They also found that 'an A3 size print viewed from approximately 8" (204mm) gives an accurate rendition of scale. It can also help to cover one eye while studying the print.' In reality they were simply increasing the focal length by reducing the distance.

A GIS for the Environmental Impact Assessment of Wind Farms by **Sparkes and Kidner** in 1996 is a case study of the spatial analysis required for an environmental statement (ES) of a proposed windfarm. It states: 'One of the main aims of an ES is to aid the public's understanding of the likely visual intrusion of the proposed wind farm. This necessitates the use of visual techniques that are both easy to understand, and yet portray relevant information.' Photomontages are described as providing 'a very accurate idea of how a new development will look in a particular location, the image is quickly understood and leaves little room for misinterpretation'. The examples given are of single frame images and as we have described in Chapter 1, this was the same period when I came across my first windfarm ES prior to the development of seamless panoramic stitching software.

In 1996 **Cumbria County Council** commissioned a study entitled *Visual Impact Study: Cumbria Wind Energy Developments,* which was undertaken by landscape architect Roger Cartwright. The study was a post construction audit of six Cumbrian windfarms approved between 1990 and 1993. The developments varied from a single domestic 6.5m turbine to 12 turbines up to 62m. Each of the six sites was viewed from up to 12 predetermined viewpoints. The landscape was filmed using an 8mm video camcorder and photographed using a 35mm camera with a fixed 50mm lens. The author concluded at Kirkby Moor that 'normal photographs taken with a 50mm lens show the turbines being just about visible, whereas in fact, they were very noticeable and the eye focused on them even more because of their movement'. He goes on to suggest that 'video film of turbines, using a zoom lens, gives a more accurate impression of the real visual impact', and can more accurately simulate the way the human eye focuses on a selected object.

Land Use Consultants were commissioned in 2000 by the UK Department of Trade and Industry (DTI) to undertake a study into the *Wind Information Needs for Planners* (2001). The report recognises the difficulties faced by planners and decision makers in accessing appropriate, complete and accurate information concerning the potential nature and effects of wind energy schemes. Ten planning authorities across the UK were selected for detailed study including consultation with planning officers and inquiry inspectors. Among the specific information gaps identified in consultation with planners and local authorities was the preparation of photomontages to represent visual impact. The reason given is 'concern that wind farm photomontages are often a poor representation of reality. This makes it difficult to assess the visual impact of schemes. It also makes it difficult to compare schemes when cumulative impacts are being considered.' The study also clearly identifies the range of the key interested parties within the planning system: planners, committee members, developers/applicants, statutory and non-statutory consultees, Reporters/Inspectors and the general public. In the final recommendation it concludes: 'We consider that a major finding of this research is the perception that much of the information which is currently available has been supplied by the industry or its representatives and is therefore biased in favour of permitting wind schemes.' They recommended that the DTI should produce Good Practice Guidance to fill the information gaps revealed.

In 2001 **Kay Hawkins and Dr Phil Marsh**, who were both members of the SNH steering group which oversaw the creation of the SNH Guidance (2006), produced a presentation for the BWEA entitled ***The Camera Never Lies***. It provides a very comprehensive explanation of the photographic process, perception, the production of fair images and their limitations. Like many more recent studies, it is rather technical and pre-dates high-resolution digital photography. However, the authors recognise that in producing visualisations 'there is an opportunity to provide each audience with the format that is most suitable for them'. The paper outlines in some detail the pros and cons of using the photographic process to replicate what we see, and observes that for the general public and decision makers, a 70mm focal length will capture the detail perceived by the eye, while acknowledging that the narrower field of view will provide less context. An edited version of this paper forms part of the draft Good Practice Guidance handed over by the University of Newcastle team in 2004 when they withdrew from their contract with SNH following the death of John Benson. The title *The Camera Never Lies* is correct in terms of linear perspective but it is my experience that the camera can certainly be used to deceive.

In 2002 a short unpublished technical paper, **Photography and Presentation of Photographs for Environmental Assessment Work: Background Information. Ian McAulay/TJP Envision Ltd. 2002** provided a significant part of the technical information in the SNH Guidance (2006), albeit with several important omissions. The paper was authored by Ian McAuley who co-authored the Guidance on behalf of Envision 3D and it provides a clear explanation of photography, lenses and perspective distances or viewing distances. It is significant that the viewing distance formula is much simpler and much of the information is founded on single frame images, but it is somewhat dated because film negative is now rarely used. This paper is not readily available.

In 2003, following the publication of the University of Newcastle Report, **The Highland Council** issued their ***Interim Recommendations for Visual Impact Assessment of Windfarms***, which was prepared in co-operation with SNH. Their stated aim was to produce consistent methodology and appraisal techniques for presentation to the council when they were considering planning applications for wind turbines. A second draft of this document has only recently come to light. The recommendations outlined are clear and comprehensive and generally based on the findings of Professor Benson. There is a requirement 'that photographs and photomontages are derived using both a 50mm and 70mm lens in a 35mm format (or equivalent combinations in other camera formats), unless otherwise justified and agreed with The Highland Council'. The points made under the heading 'Photomontages' are very pertinent to the current controversy and include a requirement that the limitations of photomontages should be recognised, particularly their tendency to underestimate the actual appearance of a windfarm in the landscape and misrepresent the relationship between vertical height and 'depth' of view; that a viewing distance of 30–50cm should dictate the technical detail of photomontage production, and that an image height of approximately 20cm should be submitted to provide improved realism. The authors certainly understood both the problems and the prescriptions required and there is some evidence that there was an improvement in visualisations for a short period during 2003 and 2004 when these recommendations were issued to applicants.

During 2004, **Perth and Kinross Council** also developed ***Supplementary Planning Guidance for Wind Energy Proposals in Perth and Kinross,*** which following public consultation was approved in 2005. Guideline 1 clearly advocates the use of single frame pictures in a paragraph on visualisations relating to the evaluation of effects, and it states that visualisations such as photomontages used to assist in the assessment of development should use a full image size of A4 or A3 for a single frame picture, giving a height of approximately 20cm to give a realistic impression. There is no evidence that these requirements were ever enforced and the enlightened efforts of both these northern councils with development management responsibility for most of Scotland's iconic landscapes were set aside by the advent of the SNH Guidance (2006).

The Symonds Report specifically addresses visual impact assessment of offshore windfarms. Although the wide and often more distant seascapes present some different challenges, the basic rudiments of realistic photomontage visualisation are, in my view, exactly the same as for onshore developments. This report, prepared by **The Symonds Group Ltd** and published in 2004, is entitled *Studies to Inform Advice on Offshore Renewable Energy Developments: Visual Perception versus Photomontage* and was commissioned by the Countryside Council for Wales. The study is a visual audit of the North Hoyle Offshore Wind Farm in North Wales, comparing the photomontages with visual perception of the constructed scheme in the field. It reaches some interesting conclusions. Although the panoramic environmental statement images were found to be technically accurate, the size of the images when compared with reality were found to be up to 50% smaller and the field of view was found to be far greater than that perceived by the human eye, which was felt to be approximately 45°. A more accurate impression of the perceived view was recorded using a 70 or 80mm focal length. The report goes on to recommend that 'photomontages should be prepared using a 70 or 80mm focal length lens and printed at a height of 20cm'. The summary concludes, 'photomontages have their limitations and … current best practice guidelines for the preparation of visualisation is misleading.'

Guidance on the Assessment of the Impact of Offshore Wind Farms: Seascape and Visual Impact Report **– DTI (now DBERR) in association with Landscape Access Recreation, The Countryside Council for Wales and SNH** was issued in 2005 and it draws heavily on the above Symonds Report to provide advice on Seascape Visual Impact Assessment (SVIA). It provides detailed advice on both photography and photomontage production, and in the preamble on photomontage its principal purpose is defined as 'a tool to assist the decision making authority in visualising how a development project, in this case offshore windfarms, would look in real life. As such it is important to provide as realistic impression as possible of the offshore windfarm as it will be seen from the viewpoint.' The authors recognise that there is a debate as to how this can be achieved and the guidance 'aims to go some way to laying to rest some of the misunderstandings in the debate about photomontages by providing clear, unambiguous recommendations and explanations of the limitations of photomontages'. It lays out the following recommendations:

• Present photomontages with an angle of view of 45–50°. Images prepared using a 35mm camera with 50mm lens (or digital equivalent) have a 40–45° field of view and are considered to most closely approximate to the central cone of vision seen by the human eye. Additional visual information beyond this angle is useful to provide context.

• Photographs should be prepared using a 35mm SLR camera with a fixed 50mm or 80mm lens (or digital equivalent), depending on the requirements and nature of the project. An 80mm lens will capture more detail and may be more appropriate for projects located at long distances from the shore.

• In order to present context outside of the central cone of view, the preferred solution when using static photography is therefore to show a laterally extended compilation merged from more than one photo frame.

• Photomontages with 40–45° field of view presented using a natural viewing distance of approximately 45–50cm, with the photomontage image printed to a height of approximately 20–24cm.

This Guidance also gives detailed consideration to virtual reality, videomontage and animated use of panoramas which can be viewed in a similar way through an on-screen viewer. Overall it is enlightened for its time and has been largely ignored. Whilst SNH were co-authors of this report, it directly contradicts the later SNH Guidance (2006) in fully recognising the role for single frames, the use of an 80mm focal length, the need for more natural viewing distances and the importance of increasing the image height. Some argue that offshore developments present a separate case; however, in terms of providing a visualisation which will look like real life there are no differences in offshore and onshore applications, particularly where distances are involved.

The US National Academy of Science publication *Environmental Impacts of Wind Energy Projects* published in 2007 gives an interesting insight into how things are approached across the Atlantic. They also recommend the use of a 50mm lens on a 35mm camera format and recommend large format prints of 10" × 12" (254mm × 305mm) as the minimum required for evaluating visual impacts. Interestingly, they make a special point about photography in clear worst-case scenario conditions and the avoidance of foreground objects within the images. All the examples in the publication are in single frame format. They also state: 'Creating technically accurate simulations is critically important. Simulations can be manipulated to produce images that either exaggerate or minimize the visual impacts of a proposed project. Accuracy should be checked by experts in the field of digital images. Another check is to have at least two independent parties provide simulations from the same viewpoint … consulting technical experts and developing standards will be important.'

SNH Scoping Advice, which was issued to developers in 2008 *(Landscape and Visual Impact Scoping Issues for Wind Farm EIA, 4th draft 2008 SNH)* under the heading 'Photomontages', redefines their requirements and is similar to the advice given to SNH by the University of Newcastle Report 2002: 'The limitations of photomontage should be recognised and acknowledged in the ES. A comfortable viewing distance of between 30–50cm should dictate the technical detail of their production. In any event, the depictions should be as realistic as possible to ensure that the general public and decision-makers are suitably informed. A full image size of A4, A3 or even greater for a single frame picture, giving an image height of approximately 20cm (min.14cm) is required to give a realistic impression.'

In January 2009, **Jim Mackinnon, Chief Planning Officer for Scotland** issued a *Letter to all Heads of Planning* in Scotland, which highlights some aspects of the problem and the importance of good visualisations. This letter was the response to a meeting held between a number of local authorities, national parks and the Scottish Government in August 2008 to consider the 'Case for Change' in visualisation practice. The Chief Planning Officer states that he is 'particularly concerned that when visualisation are used by elected members, the public and others, including planners, who are not specialists in this field, they understand how accurately the image represents what they would actually see if it were constructed'. The letter recognises that single frame images have a role to play and emphasises four points: 'visual information should be presented in a way which communicates as realistically as possible the actual visual impact of the proposal and that the format of images and the focal length of the lens will have to be taken into consideration'; all images should include the required viewing distance between the eye and image, whether single frame or a composite panorama, where it is desirable to curve the image or pan across it with the eye remaining at the recommended distance. The information given about viewing distances and their application is perhaps not fully understood by the author of the letter. The third point concerns the inclusion of viewpoint maps so that images can be used accurately on site. However, of some importance is the final point of emphasis: 'Planning authorities should make clear in any scoping advice their visualisation requirements and where these have not been provided use their powers to request further information from applicants.' The contents of this letter are now used in the Scottish Government Energy Consents Unit Scoping Requirements issued to applicants for onshore windfarm applications over 50MW installed capacity.

The **Landscape Institute Advice Note (01/09)** on *The Use of Photography and Photomontage in Landscape and Visual Assessment* was also issued in January 2009 to all members of the Institute to 'encourage the use of methods which achieve acceptable levels of accuracy, replicability, transparency of process and openness to scrutiny'. This is further emphasised in some General Principles, which state that visual representations should 'be based on a transparent, structured and replicable procedure, so that others can test and confirm the accuracy of what has been presented and thus establish trust', and that visualisations should 'be easily understood by the public'. At first reading these principles are reassuring, but the rest of the Note would appear to both contradict and impede these intentions with some confusing technical statements which fail to provide for a standard camera format or focal length. This lack of prescription makes the verification or 'scrutiny' by others almost impossible. Overall this Note failed to take account of the findings of other studies or the obvious difficulties experienced by the public audience within the modern UK planning system. This Advice Note was superseded by a further Advice Note 01/11, *Photography and Photomontage in Landscape and Visual Impact Assessment* published in March, 2011 which is included in Chapter 2.

In March 2009 SNH held a post-consultation workshop for their publication 'Designing Windfarms in the Landscape'. A presentation was given by a leading landscape architect based on his studies of windfarms in Britain and the USA entitled *Layout, Design and Visual Impact Assessment of Wind Farms: Have Landscape Professionals Got It Wrong?* **(Turnbull, 2009).** The presenter's observations of a number of windfarms in the UK and USA, which also included interviews with members of the public, revealed that windfarms can be visible from significantly further than the distances suggested in visual assessments and that visualisations commonly underplay visual impact. He posed a research question: 'Could existing methodologies used by the Landscape profession in assessing the visual impact of wind farms be under representing the actual visual impact?', and observed that the significance of effects on local people is often underestimated and that changing weather conditions have a major effect on visibility. His concluding slide states: 'The public are correct, visual impact assessments have under represented the overall significance of the effects on visual amenity as a result of a combination of factors.' The slide frames from this presentation were still available on the SNH site at the time of publication.

At the end of 2010, the **New Zealand Institute of Landscape Architects** Education Foundation published their *Visual Simulations Best Practice Guide 10.2.* This Guidance document is clear and concise and draws heavily on the approach promoted in both the SNH Guidance and the UK Landscape Institute Advice 01/09 note. They do not advocate the use of any particular focal length while recognising that a 50mm focal length continues to be widely used. In the technical section on human field of view and focal lengths there is high degree of oversimplification. The statement that a 28mm lens shows a far greater portion of the primary human field of vision than a 50mm lens is contentious. While they state that single frames may capture all that is required, there is an emphasis on a 'generally accepted' horizontal field of view of 124° for which a series of frames will be required to form a panoramic image. They do helpfully suggest that the individual single frames in a panorama should be identified. Section 7, 'Presentation of Visual Simulations', clearly identifies the factors which influence the manner of presentation, namely: what is being simulated; how and by whom the information will be used; how it will be distributed and where it will be used. Image reading distance or viewing distance is described simply and clearly until the SHN calculation formula and cylindrical viewing methods are introduced. The General Principles mirror those proposed by the UK Landscape Institute, but again, the lack of any prescription or fixed photographic standards make verification by others more aspiration than actuality. This Guidance is an example of how professional advice devised in one country is uncritically accepted and goes on to shape what is then adopted elsewhere in the world. The document can be downloaded from www.nzila.co.nz

In 2011, **Riparia**, a landscape architecture practice based in Calgary in Canada, undertook a critique of the visual impact assessment (VIA) for the Heartland Transmission Project entitled *500kV Powerline Visual Impact Assessment Critique* (June 2011). It involved assessing the VIA tools and methods used in the application and looks in some detail in Section 2 at the visualisations, which in this instance were the only tools submitted to inform the applicant's VIA. The author recognises the limitations of two-dimensional imagery but also that it is essential that visualisations are 'true to life' and comments that there are many factors and technical limitations which can result in images which are internally consistent and appear normal yet fail to accurately represent the real-world visual experience. The applicant's photomontages were created using a wide-angle lens giving a visual field of 124°. The critique explores how this approach relates to the dominant characteristics of human vision perception and asks, 'Why do we think that we see detail and colour across a wider area of the visual field than the retina can actually capture?' There follows an exploration of curvilinear projection versus planar projection, image dimension ratios, segmented or stitched panoramas and the relationship of natural perspective geometry to viewing distances along with the inherent physical and often unnatural difficulties in using them with any accuracy. Although the report gives greater consideration to large-size images or projections at A2 or greater, it recognises that images are more likely to be viewed on a computer screen than in any other form and makes the point that once the incorrect visual impression is well formed and reinforced by repetition, it is likely that they will come to shape expectations and judgements. A short section on verification makes some constructive suggestions and rightly points out that although the geometric methodology for placing proposed objects within a panoramic image is rarely the problem, in VIA it is the appropriate and perceived scaling of objects that is more crucial than accurate horizontal location. The author concludes that the applicant's methodology is a procedure for the creation of a wide-angle optical illusion of increased distance and decreased vertical object size. This fails to replicate natural perspective geometry except under non-feasible viewing conditions which are uncomfortable for most and impossible for many viewers, resulting from a disregard of visual perception, an image ratio which overemphasises the horizontal and underemphasises the vertical, and a lack of any field verification tests of the degree that the images were true to life. This is a very interesting and well-considered critique from a landscape architect with considerable experience in photography. It can be downloaded from the Riparia site at www.riparia.ca

Another recent Canadian study, entitled *A case study on visual impact assessment for wind energy development* by **Robert C. Corry**, Associate Professor of Landscape Architecture at the University of Guelph, Ontario was published in *Impact Assessment and Project Appraisal* 29(4), December 2011. This study is based on a single case study which compares the VIA visual simulations and the post-construction landscape of a 45-turbine development with a second phase of 85 turbines in flat agricultural land in Ontario. The simulations were judged for accuracy in turbine number, height, diameter and location, and adequacy in representation of the nature and extent of landscape change. The simulations were found to be only partially representative and could mislead audiences with respect to turbine number, location, ancillary attributes and to a lesser degree height. The overriding conclusion in this instance is that the single simulation frames were too narrow to adequately represent the field of view of binocular human vision. A 35mm focal length appears to have been used in the application and by the research team for initial comparisons. They then applied 120° and 360° fields of view to compare the number of visible turbines, although this is recognised to be greater than our in-focus vision but considered appropriate for a person standing in a single landscape position and turning their head. The case study references the work of Professor Stephen Sheppard and current UK guidance, with particular weight given to the SHN Guidance, to support the conclusion that in this case, a very large scheme in a flat landscape, an immersion perspective would be more representative. This case study exemplifies the difficulties of making meaningful analysis and comparison when there are no fixed photographic or image size parameters to work with and it is often difficult to follow the methodology adopted. It is of interest because it has been carried out as a way of informing future visual assessments and guidance in Ontario where there is no standard accepted methodology for preparing visual assessments. The study concentrates on turbine numbers, locations and heights in judging representativeness, but does not consider the disadvantages of wide panoramas as the sole representative image in terms of scale and distance, their practical use within a public process or alternative tailored visual solutions. The author also concludes that 'there is no evidence to suggest that the consulting professionals are affected by their commercial relationship to the client, yet an impartial consultant who creates the simulation based on developer supplied data would diminish any perception of potential conflicts of interest.' The document can be sourced through www.ingentaconnect.com

APPENDIX 7

SIZE-CONSTANCY PHOTOGRAPH

Fig 167. An enlargement of the photograph in Fig. 34 on page 29. Although the more distant person with the red top appears to be a different size when placed in the foreground, try measuring the two figures. They are exactly the same height.

BIBLIOGRAPHY

20th Century Stereo Viewers, 60 years of View-master History from Sawyers to Mattel. www.viewmaster.co.uk/htm/history.asp

Architech, 2008. *Nigg Windfarm: Review of Photomontage Visualisations Part 1*, commissioned by The Highland Council.

Architech, 2008. *Review of Photomontage Visualisations for Batsworthy Cross, Cross Moor, Three Moors and Bickham Moor Windfarm Proposals,* commissioned by the Devon branch of CPRE, Two Moors Campagn and The Exmoor Society.

Architech, 2009. *Post Construction Photographic Study of Drumderg and Green Knowes Wind Farms,* commissioned by Perth and Kinross Council.

Benson, J. F., 2005. The Visualization of Windfarms, chapter paper in I. Bishop and E. Lange (eds), *Visualization in Landscape and Environmental Planning*. Oxford: Taylor and Francis.

Bishop, I. and Lange, E. (eds), 2005. *Visualization in Landscape and Environmental Planning*. Oxford: Taylor and Francis.

Breedlove, S. M., Rosenweig, M. R., and Watson N. V., 2007. *Biological Psychology: An Introduction to Behavioural, Cognitive and Clinical Neurosciences*. 5th edn. Sunderland, MA: Sinauer Associates Inc.

Buswell, G. T., 1935. *How People look at Pictures: A Study of the Psychology of Perception in Art*. Chicago, University of Chicago Press.

Cambridge in Colour – a learning community for photographers. 2012. *Panoramic Image Projections*. Beverley Hills, California, USA. http://www.cambridgeincolour.com/tutorials/image-projections.htm

Cartwright. R. N., 1996. *Visual Impact Study: Cumbria Wind Energy Developments*. Cumbria County Council.

Corry, R. C., 2011. A case study on visual impact assessment for wind energy development. *Impact Assessment and Project Appraisal*, Issue 29(4), pp. 303–15

Countryside Commission, 1991. *Wind Energy Developments and the Landscape*. Countryside Commission publication (CCP 357).

Department of Trade and Industry (DTI) (now Department of Business, Enterprise and Regulatory Reform (DBERR)), 2005. *Guidance on the Assessment of the Impact of Offshore Wind Farms: Seascape and Visual Impact Report*. http://www.berr.gov.uk/files/file22852.pdf

Devine-Wright, P., 2005–9. *Beyond Nimbyism Project Report*. University of Manchester. http://www.esrc.ac.uk/my-esrc/Grants/RES-152-25-1008-A/read

Elert. G., 2008. *The Physics Factbook, Focal Length of a Human Eye*. www.hypertextbook.com

Ellis, H., 2008. *Planning for the End of Democracy*. Joint Planning Law Oxford Conference Paper for Friends of the Earth.

Environmental Impact Assessment (Scotland) Regulations, 1999. Scottish Planning Series, Planning Circular 8, November 2007. Crown Copyright.

European EIA Directive: 85/337/EEC & Amendments 97/11/EC.

European Directive 2003/35/EC – Public Participation Directive.

European Landscape Convention, aka The Florence Convention (Council of Europe Treaty Series no: 176) 2004.

Gregory, R.,1998. *Eye and Brain: The Psychology of Seeing*. 5th edn. Oxford: Oxford University Press.

Hawkins, K. and Marsh, P., 2001. 'The Camera Never Lies'. A paper presented at BWEA 23 Conference in Brighton, UK.

Henkel, R., 1994–2003. *StereoVision – The Cyclopean View*. Centre for Cognitive Science, University of Bremen, Germany. http://axon.physik.uni-bremen.de/research/stereo/Cyclops/index.html

Henson, D. B., 2000. *Visual Fields*. 2nd edn. Oxford: Butterworth-Heinemann.

Highland Council, 2003. *Interim Recommendations for Visual Impact Assessment of Windfarms*.

Highland Council Planning and Development Service, 2010. *Visualisation Standards for Wind Energy Developments*.

Horner + Maclennan and Envision, 2005. *Visual Analysis of Windfarms: Good Practice Guidance. Consultation Draft.*

Horner + Maclennan and Envision, 2006. *Visual Representation of Windfarms: Good Practice Guidance*. Report No. FO3AA308/2, SNH Publications.

Hunter, P. D. and Livingstone, D., 2012. *The Effect of Focal Length on Perceptions of Scale and Depth in Landscape Photographs. Implications for Visualisation Standards for Wind Energy Developments*. University of Stirling, Institute of Biological and Environmental Science commissioned by The Highland Council.

Kraft, R. N. and Green, J. S., 1989. Distance perception as a function of photographic area of view. *Perception and Psychophysics*, Issue 45(4), pp. 459–66.

Krantz, J. H., 1999. Size Constancy in a Photograph. University of Hanover, Germany. http://psych.hanover.edu/Krantz/sizeconstancy/index.html

Landscape Institute and Institute of Environmental Management and Assessment, 2002. *Guidelines for Landscape and Visual Impact Assessment*. 2nd edn. London and New York: Spon Press.

Landscape Institute, 2009. *Landscape Advice Note 01/09 – Use of Photography and Photomontage in Landscape and Visual Assessment*.

Landscape Institute, 2011. *Landscape Advice Note 01/11 – Photography and Photomontage in Landscape and Visual Impact Assessment*. www.landscapeinstitute.org

Land Use Consultants, 2000. *Wind Information Needs for Planners,* ETSU commissioned report W/14/00564/REP. DTI/Pub URN 01/779.

Lange, E. and Hedl-Lange, S., 2005. Combining a Participatory Planning Approach with a Virtual Landscape Model for the Siting of Wind Turbines. *Journal of Environmental Planning and Management,* 48(6), pp. 833–52.

Lynch, K., 1960. *The Image of the City*. Cambridge, MA: MIT Press.

Macalester College, St Paul, Minnesota, 2009. *Wind Energy: Visual Impacts and Public Perceptions*. A research project funded through the National Science Foundation. www.macalester.edu/windvisual/index.html

Macaulay Land Use Research Institute, 2010. *Visual Impact Assessment – Task Three: VIA Review*. Aberdeen, Scotland. www.macaulay.ac.uk/ccw/task-three/via.html.

Macdonald, A., 2007. *The Visual Issue: An investigation into the techniques and methodology used in windfarm computer visualisations*. Internet paper.

Mackinnon, J. (Chief Planner and Director of the Department for the Built Environment in the Scottish Government), 2009. *Letter to all Heads of Planning in Scotland*.

McAulay. I/TJP Envision Limited, 2002. *Photography and Presentation of Photographs for Environmental Assessment Work*. Version 1.2a. (unpublished).

Maguire. E. A et al.,2000. Navigation-related structural change in the hippocampi of taxi drivers. Proceedings of the National Academy of Sciences of the United States of America (PNAS). April 11, 2000, Vol. 97, No. 8 pp 4398-4403. www.pnas.org/content/97/8/4398.long

Maguire. E. A et al., 2006. London Taxi Drivers and Bus Drivers: A Structural MRI and Neurophsychological Analysis. Hippocampus 16 (12) pp 1091-110. www.fil.ion.ucl.ac.uk/Maguire/Maguire2006.pdf

Metzger, P., 2007. *The Art of Perspective*. Cincinnati: North Light Books.

National Environmental Policy Act (NEPA), USA 1969.

New Zealand Institute of Landscape Architects Education Foundation, 2010. *Best Practice Guide, Visual Simulations BPG 10.2*, www.nzila.co.nz/media/53263/vissim_bpg102_lowfinal.pdf

Nodine, C. F., Locher, P. L. and Krupinski, E. A., 1991. The role of formal art training on perception and aesthetic judgement of art composition. *Leonardo*, Issue 26, pp. 219–27.

Perth and Kinross Council. 2005, *Supplementary Planning Guidance for Wind Energy Proposals in Perth and Kinross.*

Phadke, R., 2010. Steel Forests or Smoke Stacks: the politics of visualisation in the Cape Wind controversy. *Environmental Politics*, 19(1), February, pp. 1–20.

Pirenne, M. H., 1967. *Vision and the Eye.* 2nd edn. London: Chapman & Hall Ltd.

Pirenne, M. H., 1970. *Optics, Painting and Photography.* Cambridge: Cambridge University Press.

Planning Inspectorate, Appeal Decision, Hall Farm, Routh (APP/E2001/A/07/2050015).

Planning (Environmental Impact Assessment) Regulations (Northern Ireland) 1999 and amendments.

Ray, S. F., 2002. *Applied Photographic Optics.* 3rd edn. Oxford: Focal Press.

Riparia Limited, 2011. *500kV Powerline Visual Impact Assessment Critique.* Calgary, Alberta, Canada. www.riparia.ca

Rochester Institute of Technology, Chester F. Carlson Center for Imaging Science. *Visual Perception Laboratory.* http://www.cis.rit.edu/vpl/

Rochester Institute of Technology, Chester F. Carlson Center for Imaging Science, Visual Perception Laboratory. 2004. *Eye Movements.* Rochester, NY. www.cis.rit.edu/vpl/eye_movements.html

Salvaggio, N., Stroebel, L. and Zakia, R. D., 2008. *Basic Photographic Materials and Processes*, 3rd edn. Focal Press.

Scottish Natural Heritage, 2001. *Guidelines on the Environmental Impacts of Windfarms and Small Scale Hydroelectric Schemes.* SNH Publications.

Scottish Natural Heritage, 2005. *Visual Analysis of Windfarms: Good Practice Guidance – Consultation Comments.* www.snh.org.uk/pdfs/strategy/renewable/tableofcomments.pdf

Scottish Natural Heritage, Horner + Maclennan and Envision, 2006. *Visual Representation of Windfarms: Good Practice Guidance* Report No: FO3AA308/2, SNH Publications.

Scottish Natural Heritage, Horner + Maclennan and Envision, 2005. *Visual Analysis of Windfarms: Good Practice Guidance.* Consultation Draft.

Scottish Natural Heritage, 2008. *Landscape and Visual Impact Scoping Issues for Wind Farm EIA.* 4th Draft.

Sheppard, S. R. J., 1989. *Visual Simulation: A User's Guide for Architects, Engineers and Planners.* New York: Van Nostrand Reinhold.

Sheppard, S. R. J., 2005. Validity, Reliability and Ethics in Visualizations, chapter paper in I. Bishop and E. Lange (eds), *Visualization in Landscape and Environmental Planning*. Oxford: Taylor and Francis.

Shuttleworth. S., 1980. The use of photographs as an environmental presentation medium in landscape studies. *Journal of Environmental Management,* Issue 11. pp. 61–76.

Sparkes, A. and Kidner, D., 1996. *A GIS for the Environmental Impact Assessment of Wind Farms.* Department of Computer Studies, University of Glamorgan. http://proceedings.esri.com/library/userconf/europroc96/PAPERS/PN26/PN26F.HTM

Steadman, P., 2002. *Vermeer's Camera.* Oxford: Oxford University Press.

Stevenson, R. and Griffiths, S., 1994. *The Visual Impact of Windfarms, Lessons from the UK Experience.* ETSU / W/13/00395/REP.

Stroebel, L., Perspective: Image, Camera, Distance, Objects. J. Rank Encyclopedia. http://encyclopedia.jrank.org/articles/pages/1240/Perspective.html

Sumanas Inc. Multimedia Development Service. *Visual Pathways – an animated tutorial.* Pasadena CA. USA. www.sumanasinc.com/webcontent/animations/content/visualpathways.html

Symonds Group Limited, 2004. *Studies to Inform Advice on Offshore Renewable Energy Developments: Visual Perception versus Photomontage.* Commissioned by the Countryside Council for Wales. www.socme.org/may06downloads/Photomontages.pdf

Town and Country Planning (Development Management Procedure) (Scotland) Regulations 2008.

Town and Country Planning (Environmental Impact Assessment) (England and Wales) Regulations 1999 (SI 1999 no.293) and Amendment Regulations 2008.

Town and Country Planning Environmental Impact Assessment (Amendment) (Wales) Regulations 2008.

Turnbull Jeffrey Partnership, 1995, *Photography and Presentation of Photographs for Environmental Assessment Work.* (unpublished): Background Information.

Turnbull, M., 2009. *Layout, Design and Visual Impact Assessment of Wind Farms: Have Landscape Professionals Got It Wrong? –* a presentation given at a Scottish Natural Heritage post-consultation workshop for their publication 'Designing Windfarms in the Landscape' (March 2009).

Turnbull, M., 2011. Scottish Natural Heritage, *Visual Representation of Windfarms Good Practice Guidance Review, 2011–12, Scoping Workshop 7th November, 2011.* Notes of the Workshop prepared for the Landscape Institute Technical Committee.

University of Newcastle, 2002. *Visual Assessment of Windfarms: Best Practice.* Scottish Natural Heritage Commissioned Report F01AA303A.

University of Newcastle, 2004. *Good Practice Guidance for Visual Impact Assessment of Windfarms.* SNH AB (AA308)030487. Handover Draft.

US Department of Transportation, Federal Highway Administration, Office of Environmental Policy, 1988. *Visual Impact Assessment for Highway Projects – US FHWA-HI-88-054.*

US Federal Aviation Administration, Pilot Safety Brochure, *Pilot Vision* www.faa.gov/pilots/safety/pilotsafetybrochures/media/Pilot_Vision.pdf

US National Academy of Science (Board on Environmental Studies and Toxicology, Division on Earth and Life Studies), 2007. *Environmental Impacts of Wind Energy Projects.* The National Academies Press. www.nap.edu/openbook.php?record_id=11935&page=R1

Wainwright, A., 2005–8. *A Pictorial Guide to the Lakeland Fells* (6 vols). 2nd edn. London: Frances Lincoln Limited.

Wright, L., 1983. *Perspective in Perspective.* London: Routledge.

Software: Adobe – Photoshop
Autodesk – 3ds Max
Resoft – WindFarm
GL Garrard Hassan – GH Windfarmer
EMD International, Aalborg, Denmark – WindPRO

INDEX